Sydney Barber Josiah Skertchly

Memoirs of the Geological Survey

On the Manufacture of Gun-Flints, the Methods of Excavating for Flint, the

Age of Paleolithic Man and the Connection Between Neolithic Art and the

Gun-Flint Trade

Sydney Barber Josiah Skertchly

Memoirs of the Geological Survey
*On the Manufacture of Gun-Flints, the Methods of Excavating for Flint, the Age of
Paleolithic Man and the Connection Between Neolithic Art and the Gun-Flint Trade*

ISBN/EAN: 9783337048600

Printed in Europe, USA, Canada, Australia, Japan

Cover: Foto ©berggeist007 / pixelio.de

More available books at **www.hansebooks.com**

MEMOIRS OF THE GEOLOGICAL SURVEY.

ENGLAND AND WALES.

ON THE

MANUFACTURE OF GUN-FLINTS,
THE METHODS OF EXCAVATING FOR FLINT,
THE AGE OF PALÆOLITHIC MAN,

AND

THE CONNEXION BETWEEN NEOLITHIC ART AND THE GUN-FLINT TRADE.

BY

SYDNEY B. J. SKERTCHLY, F.G.S.

PUBLISHED BY ORDER OF THE LORDS COMMISSIONERS OF HER MAJESTY'S TREASURY.

LONDON:
PRINTED FOR HER MAJESTY'S STATIONERY OFFICE,
AND SOLD BY
LONGMAN & Co., Paternoster Row; TRÜBNER & Co., Ludgate Hill;
LETTS & SON, 33, King William Street;
EDWARD STANFORD, 55, Charing Cross; and J. WYLD, 12, Charing Cross:
ALSO BY
Messrs. JOHNSTON, 4, St. Andrew Square, Edinburgh;
HODGES, FOSTER, & Co., 104, Grafton Street, and A. THOM,
Abbey Street, Dublin.

1879.

Price Seventeen Shillings and Sixpence.

LIST OF GEOLOGICAL MAPS, SECTIONS, AND PUBLICATIONS OF THE GEOLOGICAL SURVEY OF THE UNITED KINGDOM.

THE Maps are those of the Ordnance Survey, geologically coloured by the Geological Survey of Great Britain and Ireland under the Superintendence of Prof. A. C. RAMSAY, LL.D., F.R.S., &c., Director-General. The various Formations are traced and coloured in all their Subdivisions.

ENGLAND AND WALES.—(Scale one-inch to a mile.)

Maps, Nos. 3 to 41, 44, 64, price 8s. 6d. each, with the exceptions of 2, 10, 23, 24, 27, 28, 29, 32, 38, 39, 58, 4s. each.

Sheets divided into four quarters, 42, 43, 45, 46, (46 SE,) 52, 53, 54, 55, 56, 57, (59 NE, SE), 60, 61, 62, 63, 64, 71, 72, 73, 74, 75, (76 NS), (77 N), 78, 79, 80, 81, 82, 87, 88, 89, 105 (90 SE, NE,) (91 SW, NW, 93 SW, NW), (98 NE, SE, SW), (101 SE), (109 SE). Price 3s. Except (57 NW), 76 (N), (77 NE). Price 1s. 6d.

SCOTLAND.—Maps 2, 3, 7, 14, 15, 22, 24, 31, 32, 33, 34, 40, 41, 6s. each. Maps 1, 13, 4s.

IRELAND.—Maps 21, 28, 29, 36, 37, 47, 48, 49, 50, 59, 60, 61, 70, 71, 72, 74, 75, 78 to 92, and from 95 to 205, price 3s. each, with the exception of 38, 60, 72, 82, 122, 131, 140, 150, 159, 160, 170, 180, 181, 182, 189, 190, 196, 197, 202, 203, 204, 205, price 1s. 6d. each.

HORIZONTAL SECTIONS, *Illustrative of the Geological Maps.*

1 to 120, England, price 5s. each. 1 to 6, Scotland, price 5s. each. 1 to 24, Ireland, price 5s. each.

VERTICAL SECTIONS, *Illustrative of Horizontal Sections and Maps.*

1 to 62, England, price 3s. 6d. each. 1, Ireland, price 3s. 6d. 1 to 5, Scotland, price 3s. 6d.

COMPLETED COUNTIES OF ENGLAND AND WALES, on a Scale of one-inch to a Mile.

The sheets marked * have Descriptive Memoirs.
Those marked † are illustrated by General Memoirs.

ANGLESEY,—sheets 77 (N), 78. Horizontal Sections, sheet 40.
BEDFORDSHIRE,—sheets 46 (NW, NE, SW†, & SE†), 52 (NW, NE, SW, & SE).
BERKSHIRE,—sheets 7*, 8†, 12*, 13*, 34*. 45 (SW*). Horizontal Sections, sheets 59, 71, 72, 80).
BRECKNOCKSHIRE,—sheets 36, 41, 42, 56 (NW & SW), 57 (NE & SE). Horizontal Sections, sheets 4, 5, 6, 11; and Vertical Sections, sheets 4 and 10.
BUCKINGHAMSHIRE,—7* 13* 45* (NE, SE), 46 (NW, SW†), 52 (SW). Horizontal Sections, 74, 79.
CAERMARTHENSHIRE, 37, 38, 40, 41, 42 (NW & SW), 56 (SW), 57 (SW & SE). Horizontal Sections 2, 3, 4, 7, 8, 9 ; and Vertical Sections 3, 4, 5, 6, 13, 14.
CAERNARVONSHIRE,—74 (NW), 75, 76, 77 (N), 78, 79 (NW & SW). Horizontal Section 28, 31, 40.
CARDIGANSHIRE,—40, 41, 56 (NW), 57, 58, 59 (SE), 60 (SW). Horizontal Sections 4, 5, 6.
CHESHIRE,—73 (NE & NW), 79 (NE & SE), 80, 81 (NW* & SW*), 88 (SW). Horizontal Sections 18, 43, 44, 60, 64, 65, 67, 76.
CORNWALL,—24† 25†, 26†, 29†, 30†, 31†, 32†, & 33†.
DENBIGH,—73 (NW), 74, 75 (NE), 78 (NE & SE), 79 (NW, SW, & SE), 80 (SW). Horizontal Sections 31, 35, 38, 39, 43, 44, and Vertical Sections, sheet 24.
DERBYSHIRE,—62 (NE), 63 (NW), 71 (NW, SW, & SE), 72 (NE, SE), 81, 82, 88 (SW, SE)). Horizontal Sections 18, 46, 60, 61, 69, 70.
DEVONSHIRE,—20†, 21†, 22†, 23†, 24†. 25†, 26†, & 27†.. Horizontal Sections, sheet 19.
† The Geology of the Counties of Cornwall and Devon is fully illustrated by Sir H. De la Beche's "Report." 8vo 14s.
DORSETSHIRE—15, 16, 17, 18, 21, 22. Horizontal Sections, sheets 19, 20, 21, 22 56. Vertical sections, sheet 22.
FLINTSHIRE—74 (NE), 79. Horizontal Sections, sheet 43.
GLAMORGANSHIRE,—20, 36, 37, 41, & 42 (SE & SW). Horizontal Sections, sheets 7, 8, 9, 10, 11; and Vertical Sections, sheets 2, 4, 5, 6, 7, 9, 10, 47.
GLOUCESTERSHIRE,—19, 34*, 35, 43 (NE SW & SE), 44*. Horizontal Sections 12, 13, 14, 15, 59; and Vertical Sections, 7, 11, 15, 46, 47, 48, 49, 50, 51.
HAMPSHIRE,—8†, 9, 10*, 11, 12*, 14, 15, 16. Horizontal Section, sheet 80.
HEREFORDSHIRE,—42 (NE & SE), 43, 55, 56 (NE & SE). Horizontal Sections 5, 13, 27, 30, 34; and Vertical Sections, sheet 15.
KENT,—1† (SW & SE), 2† 3† 4*, 5, 6†. Horizontal Sections, sheets 77 and 78.
MERIONETHSHIRE,—59 (NE & SE), 60 (NW), 74, 75 (NE & SE). Horizontal Sections, sheets 26, 28, 29, 31, 32, 35, 37, 38, 39.
MIDDLESEX,—1† (NW & SW), 7*, 8†. Horizontal Sections, sheet 79.
MONMOUTHSHIRE,—35, 36, 42 (SE & NE), 43 (SW). Horizontal Sections, sheets 5 and 12 ; and Vertical Sections, sheets 8, 9, 10, 12.
MONTGOMERYSHIRE,—56 (NW), 59 (NE & SE), 60, 74 (SW & SE). Horizontal Sections, sheets 26, 27 29 30, 32, 34, 36, 38, 38.
NORTHAMPTONSHIRE,—64, 45 (NW & NE), 46 (NW), 52 (NW, NE, & SW) 53 (NE, SW, & SE), 63 (SE), 64.
OXFORDSHIRE,—7*, 13*, 34*, 44*, 45 (SE*, SW). Horizontal Sections, 71, 72, 81, 82.
PEMBROKESHIRE,—38, 39, 40, 41, 58. Horizontal Sections, sheets 1 and 2 ; and Vertical Sections, sheets 12 and 13.
RADNORSHIRE,—42 (NW & NE), 56, 60 (SW & SE). Horizontal Sections, sheets 5, 6, 27.
RUTLANDSHIRE,—this county is included in sheet 64.
SHROPSHIRE,—55 (NW, NE), 56 (NE), 60 (NE, SE), 61, 62 (NW), 73 74 (NE, SE). Horizontal Sections, sheets 24, 25, 30, 33, 34, 36, 41, 44, 45, 53, 54, 58 ; and Vertical Sections, sheets 23, 24.
SOMERSETSHIRE,—18, 19, 20, 21, 27, 35. Horizontal Sections, sheets 15, 16, 17, 20, 21, & 22 ; and Vertical Sections, sheets 12, 46, 47, 48, 49, 50, and 51.
STAFFORDSHIRE,—(54 NW), 55 (NE), 61 (NE, SE), 62, 63 (NW), 71 (SW), 72, 73 (NE, SE), 81 (SE, SW). Horizontal Sections 18, 23, 24, 25, 41, 42, 45, 48, 54, 57, 61, 60; and Vertical Sections, sheets 16, 17, 18, 19, 20, 21, 23, 26.
SURREY,—1 (SW†), 6†, 7*, 8†, 9. Horizontal Sections, sheets 74, 75, 76, and 79.
SUSSEX,—4*, 5, 6, 8, 9, 11. Horizontal Sections, sheets 73, 75, 76, 77, 78.
WARWICKSHIRE,—44*, 45 (NW), 53*, 54, 62 (NE, SW & SE), 63 (NW, SW, & SE). Horizontal Sections, sheets 23, 48, 49, 50, 51, 82, 83 ; and Vertical Sections, sheet 21.
WILTSHIRE,—12*, 13*, 14, 15, 18, 19, 34*, and 35. Horizontal Sections, sheets 15 and 59.
WORCESTERSHIRE,—43 (NE), 44*, 54, 55, 62 (SW & SE), 61 (SE) Horizontal Sections 13, 23, 25, 50, and 59 and Vertical Section 15.

Fig. 13.—*Flaker at Work. Boy knapping in background.*
(From a Photograph.)

MEMOIRS OF THE GEOLOGICAL SURVEY,

ENGLAND AND WALES.

ON THE

MANUFACTURE OF GUN-FLINTS,

THE METHODS OF EXCAVATING FOR FLINT,

THE AGE OF PALÆOLITHIC MAN,

AND

THE CONNEXION BETWEEN NEOLITHIC ART AND THE GUN-FLINT TRADE.

BY

SYDNEY B. J. SKERTCHLY, F.G.S.

PUBLISHED BY ORDER OF THE LORDS COMMISSIONERS OF HER MAJESTY'S TREASURY.

LONDON:
PRINTED FOR HER MAJESTY'S STATIONERY OFFICE,
AND SOLD BY
LONGMAN & Co., Paternoster Row; TRÜBNER & Co., Ludgate Hill;
LETTS & SON, 33, King William Street;
EDWARD STANFORD, 55, Charing Cross; and J. WYLD, 12, Charing Cross:
ALSO BY
Messrs. JOHNSTON, 4, St. Andrew Square, Edinburgh;
HODGES, FOSTER, & Co., 104, Grafton Street, and A. THOM,
Abbey Street, Dublin.

1879.

Price Seventeen Shillings and Sixpence.

NOTICE.

In this memoir Mr. Skertchly seems to have clearly established the circumstance that the manufacture of flint implements has been continuously carried on from what geologists have called Neolithic times down to the present day, for gun-flints are still manufactured at Brandon for the African market, and among some uncivilised tribes stone weapons are still in use, while by the Digger Indians of California stone arrow heads are made, though some years ago they preferred to make them from the thick bottoms of old porter bottles, a fact for which I am indebted to the personal observation of Mr. John Arthur Phillips.

Mr. Skertchly mentions that at Brandon the implements were the work of the inhabitants of our country before the Aryan race migrated hither, and this is likely to have generally been the case, for it is known that the Aryan races were acquainted with metals and used armour. It is, however, on record that in the Shetland islands stone knives were made and used down to quite a late period.

If it be difficult or impossible to guess with any approach to accuracy at the time when Neolithic man began to work in our area, it is still more difficult to estimate the years that have elapsed between the Palæolithic and Neolithic epochs. That man lived in this region in interglacial times I have no doubt, and I also believe it to be most probable that he even inhabited our region in pre-glacial times, and perhaps never fairly left it, but only retired south during the general increase of cold, and the gradual advance of the glaciers, and still survived in what is now the south of England. On this subject, however, Mr. Skertchly has had no occasion to enter in the present memoir.

ANDREW C. RAMSAY,
Director General.

NOTICE.

THE following Memoir, by Mr. Skertchly, gives an exhaustive and trustworthy account of an industry which seems to have been carried on at Brandon, and in its immediate neighbourhood, from a very remote period. Though formerly of considerable importance, the art has now become nearly obsolete in consequence of the improvements that have taken place in the construction of fire-arms, in the earlier history of which the present work may be considered to form an interesting chapter.

It is, also, intended to serve as an explanation of a collection of manufactured flints which has been brought together, through the aid of Mr. Skertchly, for the Museum of Practical Geology, each specimen being described and figured in the following pages.

In his treatment of the more obscure ethnological and archæological questions, the opinions of the Author are entitled to every consideration. The practical acquaintance which he has acquired of flint-knapping and working, his close study of the different ways in which flint can be broken, coupled with his intimate knowledge of the geological structure of the country around Brandon, as well as of the deposits in which Palæolithic implements have been found, necessarily impart additional weight and value to his arguments with regard to the geological age of Palæolithic man; and also to his endeavours to trace a connexion between Neolithic art and the modern flint manufacture as practised by the Brandon knappers of the present day.

The specimens of flints, and the tools and appliances used in their manufacture, have been drawn, on the wood, by Mr. Redaway, from the objects themselves, and have been engraved by Mr. Shepherd.

H. W. BRISTOW,

Senior Director.

Geological Survey Office,
    28, Jermyn Street, London, S.W.,
        23rd August 1878.

PREFACE.

In this work the manufacture of gun and other flints as carried on at Brandon, is described in greater detail than has before been attempted; and the value of the study of this branch of industry to geologists is pointed out. The volume is especially descriptive of a very complete series of specimens made under my supervision for the Museum of Practical Geology by Mr. W. J. Southwell, to whom my best thanks are due, not merely for the care he bestowed upon the specimens, but for the unflagging patience with which he imparted to me day by day most of the information herein contained concerning his craft. So far as the Brandon manufacture is concerned, I may claim to have produced a work as free as possible from errors of description; for most of it was written in the workshop, and all has been revised by my unwearied flint-knapper. Similar assistance has been rendered by stone-diggers in that branch of the work which relates to their craft. So far I have been merely the mouth-piece of the flint-workers, but have verified every part of the description by learning the business practically.

That the Brandon gun-flint manufacture is a direct descendant of the neolithic age seems to me certain from a comparison, 1, of old "scrapers" with old strike-a-lights, 2, of old strike-a-lights with modern ones, 3, of strike-a-lights with old English or modern French gun-flints. It is further shown that in one nearly obsolete tool we probably have an iron *replica* of a neolithic flaking-hammer, and that some of the curious, small, bored celts answer in every point to flaking-hammers and to nothing else.*

The observations upon the kind of hammer used, the character of the blows given, and the resulting nature of the fracture produced are original, and founded upon a close study of the different modes in which flint can be broken.

The illustrations are nearly all very faithful copies of particular specimens. Owing to my not being able to superintend the drawing, some of the gun-flints are slightly faulty, the trimmed heels being omitted. If it be remembered that all but double-edged flints have a trimmed heel, no error can arise in naming the flints. The English hammer has suffered perfection. It is (as needs must be) a very battered specimen, but the kindly artist has feelingly restored its lost beauties, and unwittingly robbed it of its more important peculiarities. It is well also to remark that the *Doubles* in this collection are all rather too large, and hence too much like the *Singles*. This will be remedied if this work attains to a second edition.

To Mr. H. W. Bristow, F.R.S., Director of the Geological Survey of England, my best thanks are due for the verification of many of the references, and much kind assistance in many ways.

SYDNEY B. J. SKERTCHLY.

Brandon, February 28th, 1876.

---

* These observations have now been largely extended, and many new "*points d'appui*" have come to light between the ancient and modern arts.

This, and the question of the age of Palæolithic Man, will be treated of in a distinct work.

S. B. J. S.

December 1877.

TABLE OF CONTENTS.

LIST OF ILLUSTRATIONS.

x

MANUFACTURE OF GUN-FLINTS.

INTRODUCTORY.

THE manufacture of gun-flints has been carried on for a great length of time at Brandon, a little town on the Ouse Parva, about six miles west of Thetford, in Suffolk. Gun-flints were also made at Icklingham, on the river Lark, in the same county, about 12 miles south of Brandon; of late years only a single shop and workman have existed at this place, and now (1875) the manufacture has been stopped for two years. At this place the manufacture never attained the magnitude or the degree of excellence which gave to Brandon the pre-eminence it now enjoys. Near Norwich gun-flints are still made in small quantities by a single workman, but the Norwich knappers being Brandon men this place may be considered a branch of the mother trade.

Mr. Evans, F.R.S., in his "Ancient Stone Implements of Great Britain,"* gives an account of the manufacture at Brandon, but the fullest record is that of Mr. James Wyatt, F.G.S., of Bedford, in Stevens's "Flint Chips."† From this admirable paper free quotation will be made respecting foreign manufactures. In some few instances the two authors named have fallen into trifling errors, which the possession of their works, and my residence at Brandon, have enabled me to rectify. I have, however, been able to treat the question in far greater detail, and from different aspects than have before been attempted. Like Mr. Wyatt, I have studied the art of gun-flint making under experienced Brandon professors.

At first pyrites was as often used as flint, and this must have continued until the manufacture of gun-flints was permanently established. Beckmann, in his "History of Inventions," seems to be of opinion that the use of pyrites preceded that of flint in all cases. This, I think, was not the case, for in many places flint is much more plentiful than pyrites; it is better adapted for producing sparks, and was, in fact, in general use as part of the old "flint and steel" apparatus. It is suggestive, on this point, that in the Tower not a single English weapon is preserved in which pyrites was used instead of flint. It must, however, be remembered that in early (neolithic) times pyrites was used with flint for obtaining fire.

---

* Pp. 16 to 20.   † Pp. 578–90.

The date of the final abandonment of pyrites for flint is not known, but in the year 1586, Julius, Duke of Brunswick, had pyrites collected for his fire-arms near Seefen, and even worked it into shape himself, "though in so doing he often bruised his " fingers, and was advised by the physicians not to expose him- " self to the sulphurous vapour emitted by that substance."

The development of fire-arms is finely illustrated by the col- lection preserved in the Tower of London, from which the following examples are described. At first the flints were merely rough pieces of convenient size, and these were used occasionally as late as the year 1630 in France and 1738 in England The crudest type of fire-arm I have seen is in this collection. It is an iron spear, about 3 feet in length, with a barrel at the butt-end, as shown in Fig. 1, which is one-half the size of the original. The barrel is 6 inches long, with a bore of about $\frac{1}{4}$ of an inch in diameter. The touch-hole and pan are shown at d. Attached to a link are two short wire chains, to one of which a stout steel pin is appended and to the other a clip to hold the flint. The weapon

Fig. 1. *Hand-gun and spear.   In the Tower Armoury.*

    a. b. Length of barrel.
    c. Portion of the spear.
    d. Touch-hole and pan.
    e. Steel pin.
    f. Clip for flint.

was placed under the arm, the flint was then held over the pan and struck with the steel pin. It is dated 16th century, and numbered $\frac{14}{5}$.

At Dresden an old gun is still preserved, called a *Büchse* " on " which, instead of a lock, there is a cock with a flint-stone " placed opposite to the touch-hole, and this flint was rubbed " with a file till it emitted a spark."[*]

The first improvements on the original hand-fired match-lock was a cock to hold the match. The cock fell towards the gunner and a sliding cover was afterwards appended to the pan to protect the priming, and prevent sparks striking the gunner's face. The earliest weapon in the Tower in which the motion is from the gunner is a Flemish wall-piece of the middle of the 17th century.

---

[*] History of Inventions, Discoveries, and Origins, by John Beckmann, London. Bohn. 4th edition, pp. 533 to 539.

It is labelled $\frac{12}{24}$. A German wheel-lock harquebus, dated 1738, however, still has the cock falling towards the gunner.

In the 15th century the wheel-lock was introduced with the use of flints or pyrites. The stone was screwed into a cock, and a steel plate or wheel, which was cocked or wound up by a particular kind of key called a spanner, was fixed to the barrel. At first these weapons missed fire very often, and a German harquebus, dated 1546, and marked $\frac{19}{5}$, was originally a wheel-lock but converted into a match-lock. Afterwards a match-lock was used with the wheel-lock so as to be ready in case of the latter missing fire. The harquebus with this double arrangement dated 1603, is exhibited at the Tower, the catalogue mark being $\frac{12}{331}$. Wheel-locks were used in Germany as late as the year 1797, for a rifle so dated is shown with the catalogue number $\frac{12}{64}$.

The date of the introduction of the true flint-lock is uncertain. The earliest notice known to Mr. J. Hewitt, the author of the "Official Catalogue of the Tower Armouries," is that printed in the 1st vol. of the proceedings of the Norfolk Archæological Society, the record of a payment by the Chamberlain of Norwich in 1588, 'to Henry Rador, smyth, for making one of the old pistols with a snapphance and a new stock for it,' (p. 16.) "The German name of Schnapphahn," says Mr. Hewitt, "borne by the flint arm in its earliest days, clearly shows that the invention was a German one."[*] The superiority of the flint over the match-lock is shown by the fact that soon after its introduction the old locks were converted into the new form, specimens of which are to be seen in the Tower, as for instance one marked $\frac{12}{157}$, of the reign of William III.

Flint-locks were introduced into the English army about the year 1686, and were in general use at the beginning of the 18th century.

The soldiers of the duchy of Brunswick used match-locks until the year 1687, "In France, the Miquelet gun-lock, a Spanish in-" vention, was first introduced in 1630, but the stone, if flint was " used at that time, had not, at all events, been subjected to any " manufacturing process."[†] By the year 1703 the soldier was armed with a musket, but he had to find his own flints which were often used in a rough state. It was not until the year 1719 that a manufacture of gun-flints was regularly established in France.

Gun-flints were superseded in England by percussion caps about the year 1835, and so complete was the change that flint-locks ceased to be manufactured for home use soon afterwards, with the exception of those used for horse-pistols, which are still very frequently of the old pattern, and which may be purchased at most ironmongers in large towns. The old flint-lock, long-barrelled, duck-guns used by gunners are by some still preferred to modern guns, for it is said, and it seems to me very likely, that the flash in the pan causes the ducks to raise their heads from the water,

---

* Official Cat. Tower Armouries, p. 74 (foot note) 1870.
† Quoted by Wyatt. *Flint Chips*, p. 579.

and so increases the chance of hitting them. The flash is certainly
sufficiently in advance of the discharge to be noticed, and wary
birds like ducks are very likely to be roused by it.

A good flint will last a gunner half a day, but as there is considerable uncertainty as to amount of work got out of it in that
time, I fired a flint-lock pistol with a new flint 100 times in
succession.* The following results were obtained :—

The pistol *fired*   -   - 36 times.
   „   „ *flashed*   -   - 25 „
   „   „ *missed fire* -   - 39 „

100

I did not find much difference between the beginning and end
of the experiment ; indeed the misses were more frequent at first
than afterwards. Thus dividing the 100 shots into four batches
of 25, in the order of discharge, there were :—

——	1.	2.	3.	4.
Fires - -	11	10	9	6
Flashes - -	1	7	9	8
Misses - -	13	8	7	11
Totals - -	25	25	25	25

When a flint gets much worn, however, it misses fire very often,
a serious matter in warfare. " All military men," says a writer in
Rees's Cyclopædia, " must know that nothing is more adverse to
" the operations of a regiment than the necessity (which too often
" occurs in consequence of the proper form of gun-flints not being
" sufficiently attended to) for men to quit their ranks for the
" purpose of either hammering or changing their flints. To brave
" men such a necessity is painful as well as dangerous, while to
" the less resolute it serves at least for a pretext to pass into the
" rear, or eventually to relinquish his post altogether."†

The gun-flint trade is steadily dying out, and unless some
alteration takes place a few more years will terminate its
existence. Its decay is not owing, as might have been supposed,
to a falling off in the demand ; but to a lack of hands, the boys
preferring agricultural or other labour to the confinement of a
knapper's shop. The demand for flints, especially from Birmingham and Sheffield merchants, is even now in excess of the supply,

---

* The flint used was rather too broad. A better fitting one fired 34 times, flashed
7 times and missed 59 times. In the first 25 it fired 20 times, flashed 1 and missed
4 times ; in the last 25 there were 2 fires and 23 misses—showing that the flint was
worn up. It may be taken as near the truth that a flint cannot be depended upon for
more than 30 shots.

† Quoted by Wyatt.

and to my knowledge large orders for hundreds of thousands of gun-flints have been recently declined in consequence of the paucity of hands. It is very difficult to obtain statistics to prove this; but old men say the number of knappers has steadily decreased. In the year 1868 there were 36 men regularly at work, exclusive of the stone diggers; these men worked for three masters who were buyers in Brandon and dealt with the merchants. Of late years the flint-makers have preferred to get rid of the buyers, and to sell directly to the merchants. There are at present (April 1878) 26 men and boys, who work in four shops, showing a decrease of 10 knappers in ten years. Several knappers work at other things during the day and knap at night, but many of these are giving up flint-making. Five men and five boys are engaged in stone-digging.

The gun-flint manufacture used to be considered a very unhealthy one, and it was said that every knapper died of consumption at about 40 years of age. This may have been true, but the consumption was of drink, and not of the lungs; it being the practice to work from Wednesday to Saturday and drink the rest of the time. This unhappy state of affairs is coming to an end, and some of the steadiest men in Brandon are knappers, and several are quite old. The practice of damping the flint before using it also renders the trade less unhealthy; but the particles of dust flying about certainly have some influence upon health.

GEOLOGICAL POSITION OF THE FLINTS.

The flint now manufactured at Brandon is obtained from Lingheath, about a mile south-south-east of the town, but it has, until of late years, been dug at Santon Downham, three miles further up the river, and at Broomhill about a mile from the town on the Norfolk side of the river. The flints occur, of course, in the Upper Chalk, and the sections at the three points will now be described.

Lingheath.—Lingheath is completely honey-combed with new and old pits, from Brandon Park on the west to the slope of the Ouse Valley on the east. The pits in the latter situation are now worked out; they were shallower than those high up on the heath, a necessary consequence of their position. They are known as the *Fleet Pits* from this circumstance, *fleet* being a local term signifying *near the surface*, as distinct from *gain* meaning near at hand in a horizontal direction; fleet refers to vertical, gain to horizontal distances. These admirable terms express two distinct ideas for which our cultured language has only the one word *near*. The flint has been worked on Lingheath for about 160 years, prior to which time, the stone was obtained from Brandon Park near the Elms.

Fig. 2 is a section of a typical flint-pit measured by myself on the summit of Lingheath, in the Poor's Plantation, to which is

added from the information of old diggers all the beds below the
Floor-stone. The section thus compounded is as follows :—

		ft.	in.
1. Sand and Gravel	- - -	3	0
2. Dead-Lime	- - -	5	0
3. Soft, White Chalk	- -	4	0
4. Horns Flint, thickness included in	-		
5. Soft, White Chalk	-	3	6
6. Toppings Flint	- -	0	5
7. Soft, White Chalk	- -	4	0
8. First Pipe-Clay	- -	0	4
9. Hard, White Chalk	- -	1	0
10. Upper-Crust Flint	- -	0	8
11. Soft, White Chalk	-	1	0
12. Second Pipe-Clay	- -	0	2
13. Hard Chalk, one jointless bed		1	0
14. Soft, White Chalk	-	2	0
15. Wall Stone	- - -	1	0
16. Very Soft Chalk, full of Horns	-	2	6
17. Soft, White Chalk	- -	2	6
18. Third Pipe-Clay	-	0	2
19. Hard, White Chalk	- -	3	0
20. Floor Stone	- - -	0	8
21. Soft, White Chalk	- -	7	6
22. Hard, White Chalk	-	1	6
23. Rough and Smooth Blacks	-	0	4
24. Soft, White Chalk	-		
		46	1

The *sand* and *gravel* is part of the very wide-spread deposit,
containing palæolithic implements, which covers almost the whole
face of the country, reaching the highest ground in the county and
plunging into the valleys quite irrespective of the present drainage
system. Mr. Evans, F.R.S.,[*] and Mr. Flower, F.G.S.,[†] have
described portions of this deposit as isolated patches capping high
ground. The latter indeed says "it comprises an area of from
thirty to forty acres," had he written *miles* he would have been
within the truth; neither of these authors seem to have a knowledge of the extent of these beds, which will be described in the
explanation of sheet 51 N.E. of the Geological Survey.

The term *dead-lime* is a local designation for decomposed chalk,
into which the sand penetrates, and with which its upper part is
generally mixed. It is quite friable above but lumpy below;
small flints, with natural coats, are scattered through it. In
"fleet pits," where it intersects a layer of flints (as will be
presently described), the large flints are always "edge-ways,"
or stand on end, and often have brown glazed coats. The junction
with the chalk below is more or less abrupt.

* Ancient Stone Implements of Great Britain, 1872, pp. 507, 511, &c.
† Q. J. Geol. Soc., vol. xxv., 1869, pp. 449-50.

1.	Sand and Gravel.
2.	Dead-lime.
3.	Soft, White Chalk.
4.	Horns Flint.
5.	Soft, White Chalk.
6.	Toppings Flint.
7.	Soft, White Chalk.
8.	First Pipe-Clay.
9.	Hard, White Chalk.
10.	Upper-Crust Flint.
11.	Soft, White Chalk.
12.	Second Pipe-Clay.
13.	Hard Chalk.
14.	Soft, White Chalk.
15.	Wall Stone.
16.	Soft Chalk, with Horns.
17.	Soft, White Chalk.
18.	Third Pipe-Clay.
19.	Hard Chalk.
20.	Floor Stone.
21.	Soft, White Chalk.
22.	Hard Chalk.
23.	Rough and Smooth Blacks.
24.	Soft, White Chalk.

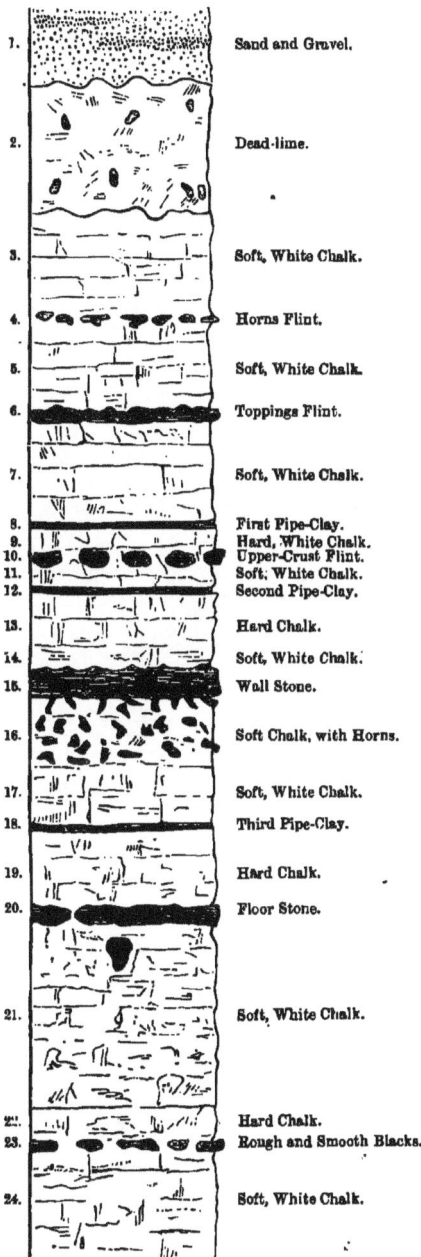

Fig. 2.—*Section of Flint Pit at Lingheath.*

The *soft white chalk* beds 3, 5, 7, 11, 14, 20, and 23 are of the ordinary character, and are said not to make such good lime as the dead-lime, which is always preferred for mortar. Bed 16 is very soft, and often stained yellow. It is sometimes mixed with sand to render it stiffer, and is shovelled out.

The *hard chalk*, beds 9 and 18, is a hard, sub-crystalline limestone, which rings and strikes fire under the strokes of the pick. The stone is so hard that it cannot be picked on the solid face, but has to be worked from the joints. No. 18 is sometimes only 2 feet thick, in which case 6 inches of soft chalk overlie the floor stone. Bed 13 has no joints. Bed 22 is of similar material, but is never worked into except along the "burrows" beneath the floor-stone; but trial pits were sunk many years ago in search of flint as shown in the section.

The *pipe-clays* are thin seams of marl, and are pretty constant, especially the lowest or third, which the workmen say "rules" the floor-stone—that is to say, when it is reached the floor-stone is known to be only 3 or 4 feet distant.

Horns is the appropriate term applied to an irregular line of flints, which are nearly all small and finger-shaped; they seldom run to more than 3 inches in length, and half-an-inch in breadth.

The *Toppings* are the first regular layer of flint, No. 6 in the section. They are more or less continuous, or form, as the workmen say, a *sase* or *sese*. They are "hobbly" stone, that is, covered with "paps" or knobs on the top, but flat below. They break "grisly," that is grittily, and do not "run," or flake cleanly, and are "coarse working stone," that is to say, from their grisly nature they do not cut clean, and will not make "best" flints. Good pieces are occasionally found, and these are not hobbly but flat like floor-stone. Toppings are nearly always "burrowed" for in filling up the pit, and are worked from beneath.

The *Upper Crust Flints*, No. 10 in the section, are generally round and lumpy, and do not form a regular *sase*, but are dotted here and there in the same line. They are nearly always grey in colour, without paps, and "double coated," that is, they have two distinct layers of cherty matter on the outside, which break away separately. These coats are sometimes parted by a thin layer of flint. Upper-crusts are only used as building-stones, and merely taken out in sinking the shaft, but when building-stones are in demand they are "burrowed" for. They are not "faced," but used rough as they are dug, and are known as *rough builders*. Occasionally, as at Elvedon, these flints are of good quality. They are there then known as "best upper-crusts."

The *Wall-stone*, No. 15 in the section, is always continuous, or forms a "sase." It has "paps" above, and horn-like projections below called "legs," which are sometimes a foot long, and make the stone very difficult to raise. The pieces come away in long flat masses sometimes a yard square. Wall-stone is generally

black but sometimes grey or spotted, and occasionally has a
bluish " plumage," whence they are aptly termed " jackdaw "
coloured.   The stone is nearly always of good quality, flakes well,
with little waste, and hence only leaves small cores for " builders."
Wall-stone is burrowed from the top, as the legs would prevent
it being worked from below.

Beneath the wall-stone the chalk is full of " horns " for about
2½ feet, so thick together that the pick can hardly be used.

The *Floor-stone*, No. 19 in the section, is the bed to which the
pits are sunk, and from which most of the gun-flints are made.   It
is generally continuous, but sometimes in ovoid masses which are
called " heel pieces," but even then the " heels " of adjacent stones
are in contact.   In some places paps are found on the top of the
flint, but these are rough, and in this respect different from the paps
of the toppings ; such stone is called " rough topped."   Another
variety has an undulating surface ; such stone being called " hobbly
topped."   These are very seldom heeled, and are easiest to get,
because when they break away they always leave a " face " to
work upon, and therefore no time is lost in picking chalk.   The
floor-stone is nearly always flat-bottomed, and is thinnest when
there are many heel-stones.   When the stone is over a foot in
thickness it is generally grey in the middle.   Very rarely the
floor-stone runs into great " harp-like " pieces 4 feet 6 inches
across, which I take to be " *Paramoudras*."*   They are so ex-
ceptional that when my informant found one, he sent for other
diggers to look at the "cūrosity."   He got a " jag " of stone from
his cu-rosity, that is, a one-horse cartload, about equal to a ton.

More commonly, but still rarely, similar stones are found just
below the floor-stone, which are described as " like gret eggs," and
from each of which half a jag of stone can be got.   They are
called " gulls."

Floor-stone is always burrowed for, and gulls too when they can
be found.

Rough-Blacks and *Smooth-Blacks* are the names applied to large
detached flints, which were found ten feet below the the floor-stone
in some trial-pits sunk many years ago.   They occur too sparingly
to be remunerative.   The smooth-blacks were some of the best
working stones ever raised ; being good in colour, clean-cutting,
and of good running quality.   The rough-blacks were grisly and
fit only for common gun-flints.   The surrounding chalk is described
as very hard.

Santon Downham.—The flint-pits, no longer worked, were
situated on Santon Downham Warren, opposite the Warren House,
and near a tumulus locality known as Blood Hill   There is
another mound on Eriswell Rabbit Warren, near High Lodge
Farm, which goes by the same name.   The pits are on the slope

---

* *Paramoudra* is a local Irish term, adopted in science, applied to large cup-shaped
flints, which are also known as pot-stones, &c.

of the valley-side, and are consequently shallow. The general
section is as follows :—

		ft.	in.
1. *Sand and Gravel* - - -		3	0
2. *Dead-lime* - - - -		5	0
3. *Third Pipe clay* - - -		*trace.*	
4. *Hard Chalk* - - - -		2	0
5. *Floor-stone* - - -	6 in. to	3	0
6. *Soft Chalk* - - -		3	0
		13	**0**

The *sand* and *hard chalk* beds call for no special notice. The
dead-lime contained a few floor-stones " edge-ways," or on end,
most of which had brown, glazed coats. The *pipe-clay* was just
distinguishable below the dead-lime. The floor-stone was rich in
" egg-shaped gulls " on *Paramoudras*, some of the stone was milky
in colour, like black flints changed by the sun. They were, however, good stone, and made good gun flints.

Broomhill.—Flint was dug formerly between the neolithic
pits known as Grime's Graves and Broomhill Plantation. The
section in the modern pits is like that of the ancient ; indeed very
little variation can be traced in the sequence of the beds anywhere
round Brandon. The pits are deep, but the chalk commences
below the horns. The following is a typical section : —

		ft.	in.
1. *Sand and Gravel* - - -		3	0
2. *Dead-lime* - - -		5	0
3. *Soft, White Chalk* - - -		3	0
4. *Toppings Flint* - - -		0	5
5. *Soft, White Chalk* - -		3	0
6. *First Pipe-clay* - - -		0	1
7. *Moderately Hard, Grisly Chalk, with red stains* - - -		3	0
8. *Upper-Crust Flint* -	2 ft. to	0	8
9. *Soft, White Chalk* - -		3	0
10. *Second Pipe-clay* - -		0	1
11. *Soft, White Chalk* - -		3	0
12. *Wall-stone* - - 1 ft. 6 in. to		1	0
13. *Soft, White Chalk, full of horns*		2	6
14. *Soft, White Chalk* - - -		2	6
16. *Third Pipe-clay* - - -		0	1
17. *Hard Chalk* - - - -		3	0
18. *Floor-Stone* - - - 3 in. to		0	4
		34	**8**

The *sand* and *gravel* contained a few palæolithic implements.
The dead-lime yielded a few brown-glazed edgeways flints. Bed

3 in this section is the same as No. 5 at Lingheath. Bed 5 is thinner at this place, and beds 7 and 9 are thicker at this place than the corresponding beds at Lingheath.

In the large ballast-pit adjoining the railway-line some very good flint was found, a rather remarkable circumstance, for the flint in gravel beds is nearly always full of cracks. This is the place where Mr. Evans described the occurrence of singular hollows, which were supposed to be old dwellings, but which he shows to be natural cavities caused by the gravel and sand sinking into pipes.* He obtained some implements from the gravel, and I have recently found one very much like that figured by him from Whitehill, Thetford, further up the river.†

Shaker's Lodge.—At Shaker's Lodge, on Wangford Rabbit Warren, about 2·5 miles south of Lingheath on the summit of the hill, floor-stone was met with in sinking a well, and a trial-pit was opened ; the bed, however, has not been worked. The section is still open and shows :—

			ft.	in.
1. Sand, full of angular flints		-	3	1
2. Dead-lime	-	- 1 ft. to	3	0
3. Hard Chalk	- -	- 3 ft. to	1	0
4. Floor-stone	-	-	0	4
			7	5

This section affords a means of calculating the dip of the beds. The surface of the ground is 163 feet above Ordnance datum, the floor-stone 158 feet. The surface at the Poor's Plantation, Lingheath, is about 50 feet above datum, the floor-stone 15 feet. The horizontal distance being 2·5 miles gives as the dip 143 feet in that distance, or 57·2 feet per mile, equal to 1 in 93 or about 0·5°.

Elms Plantation.—Old flint-pits are seen in and around Elms Plantation in Brandon Park, about a mile due north of Shaker's Lodge, but I have not been able to obtain any account of the section. They were worked out before the Lingheath pits were started, and may perhaps have been dug before the gun-flint trade arose.

Norwich.—A few gun-flints are still made at Catton, near Norwich, about 1½ miles north-east of the city. The knapper is named Frewer, and is an emigrant from Brandon. The shop is in a large pit, in which Middle-glacial sands and gravels, brick-earth and chalk are dug. The chalk is very different from that at Brandon, and is most probably not nearly so old. It is nearly all soft, in regular courses, stained of a light yellowish-brown

---

* Geol. Mag. vol. v., p. 445, 1868.
† Anc. Stone Imp. Fig. 432, p. 500.

colour, and is dug to a depth of about 25 feet, six lines of flint being passed through in that distance. The flints are scattered and never tabular; they vary in size from a few inches to a foot in diameter, and are irregular in shape, or, as they would be described at Brandon, hobbly. The coats are moderately thick and soft, and the stone is very rarely of a good black colour; every piece I examined in company with my colleague, Mr. H. B. Woodward, F.G.S., was mixed with grey and grisly. The pit is not regularly worked for flint, but the larger pieces are selected as the chalk is removed. No distinction is made between the several layers, nor could I detect any constant features. Comparatively few flakes can be struck from a quarter in consequence of the inferior quality of the stone; and the flints, for the same reason, are not so well finished.*

Icklingham.—Icklingham is a village on the R. Lark, about 3 miles east of Mildenhall. About 30 years since it was the chief seat of the gun-flint manufacture, the workmen being Brandon men who returned home every Saturday. The workshops are now either in a ruinous condition or converted into cottages, but around some of them the heaps of waste chips still remain. The cause of the temporary abandonment of Brandon by the knappers was the superior quality of the stone at Icklingham, and when difficulties arose concerning the raising, the trade reverted to Brandon, where it is still carried on more vigorously than elsewhere. For some years past only one knapper has lived in Icklingham, Henry Ashley, a Brandon man, who raised his own stone and worked it into gun-flints; but even he has done nothing for two years, and does not think of resuming his old trade, finding day-labour to pay better. Mr. W. J. Southwell was good enough to go with me to Icklingham, and Mr. Ashley, at his request, kindly supplied me with the following information.

The gun-flints lately made at Icklingham were:—

1. *Best Muskets.*
2. *Second do.*
3. *Common, or Grey do.*
4. *Best Carbine, single edge.*
5. *Second Carbine, do.*
6. *Common Carbine, do.*
7. *Best Horse-Pistol.*
8. *Second Horse-Pistol.*
9. *Common Gun.*
10. *Small Gun.*
11. *Best Seconds.*
12. *Worst Seconds.*
13. *Round Discs.*

The common gun is the same as the common horse-pistol; the

---

* Since writing this the trade has been abandoned.

small gun is the fine single, and the seconds are doubles. Specimens of these Icklingham flints are exhibited ; but, notwithstanding the superior quality of the flint, their workmanship is much inferior to that of the Brandon specimens.

The stone used at Icklingham was dug on Icklingham Heath, close to the Seven Trees, 1½ miles north of the village. The place is called Seven Trees Brick ; the first part of the term applies to a clump of elms, of which five only remain, and the latter is synonymous with *Field,* an open space. The pits were very numerous, perhaps 500 in number, but their area is circumscribed. None were open at the time of my visit, and the only noticeable feature was the close proximity of the shafts, which averaged about nine yards from one to another. The section at this place was, in descending order :—

		ft.	in.
1. *Sand* -	1 *ft.* 6 *in. to*	3	0
2. *Dead-lime* -	-	6	0
3. *Soft, White Chalk* -	-	3	0
4. *Toppings Flint* -	5 *in. to*	1	0
5. *Soft Chalk* -	-	3	0
6. *First Pipe-clay* -	- 1 *in. to*	4	0
7. *Hard Chalk* -	-	2	0
8. *Upper-Crust Flint* -	6 *in. to*	1	6
9. *Soft, White Chalk* -		3	0
10. *Second Pipe-clay* -	-	0	1½
11. *White Chalk, rather hard* -	-	3	0
12. *Wall Stone* -	1 *ft. to*	1	6
13. *Soft Chalk, full of horns* -	-	3	0
14. *Hard Chalk* -	3 *ft. to*	5	0
15. *Third Pipe-clay* -	-	0	1
17. *Floor-stone* -	- 3 *in. to*	1	0
18. *Moderately Hard Chalk*	- 2 *ft. to*	3	0
19. *Very Hard Chalk* -	- 8 *ft. to*	9	0
20. *Rough and Smooth Black Flint*	3 *in. to*	1	0
Mean about - -		45	0

The similarity of this section with those at Lingheath and Broomhill, from which it is distant 8 and 10 miles respectively, is very striking. Bed 7 is in large blocks, and rings and smokes under the strokes of the pick. The upper-crusts were large scattered flints, mostly grey. The rough and smooth blacks formed a regular layer or *sase,* in this differing from the stone at Lingheath. They were only occasionally worked, and were burrowed from above like wall-stone.

A very remarkable pit was sunk by Mr. Ashley about 300 yards west of the above pit, in which the first three "stones" were missing. The section was as follows :—

						ft.	*in.*
1.	*Sand*	-	-	-	-	3	0
2.	*Dead-lime*	-	-	-	-	6	0
3.	*Soft, White Chalk*	-	-	-	-	7	0
4.	*First Pipe-clay*	-	-	-	-	0	3
5.	*Soft, White Chalk*	-	-	-	-	0	6
6.	*Second Pipe-clay*	-	-	-	-	1	0
7.	*Soft, White Chalk*	-	-	*about*	12	0
8.	*Third Pipe-clay*	-	-	-	-	0	1
9.	*Hard Chalk*	-	-	-	-	2	0
10.	*Gulls (Paramoudra) in three tiers with* 3 *in. partings of chalk between*					9	0
11.	*Floor-Stone in a regular sase.*						

The gulls are described as being nearly 3 ft. in height, grey and grisly, and of no use for gun-flints. The thicknesses of the chalk beds in the above section must be much overstated, for Ashley told me the pit was 25 ft. deep, whereas his account makes it 40 ft. It is certain, however, that the pipe-clays were found as stated, and the local absence of the toppings, upper-crusts, and wall-stone, and the presence of gulls above the floor-stone are very singular.

The stone obtained from Seven Trees Brick is very good, and of much better average quality than that found near Brandon. It is generally intensely black, often jackdaw-coloured, of good running quality, and the coats are hard. To save expense of cartage Ashley often quartered and flaked his stone on the spot, and the numerous cores lying about show how profligate he was of his wealth of stone, for very many of them would be gladly flaked over again by the Brandon men.

Elvedon.—Elvedon is 5 miles S.E. from Brandon, and 4 miles S.W. of Thetford. Good stone is obtained from a pit close to the lodge of the Maharajah Dhuleep Singh's park. The stone is not specially worked but obtained from a chalk pit. The section is as follows :—

						ft.	in.
1.	*Sand*	-	-	-	0 *to*	3	0
2.	*White Chalk, with scattered horns flints*	-				11	0
3.	*Pipe-clay*	-	-	-	-	0	4
4.	*Upper-Crust Flints*	-	-	-	-	0	8
5.	*White Chalk*	-	-	-	-	4	6
6.	*Wall Stone*	-	-	-	-	1	0
7.	*White Chalk*	-	-	-	-	4	0
						24	6

The upper-crusts here yield the good flint. This is unusual, but sometimes occurs at Lingheath; the stone is known as "best upper-crust."

B

Elvedon Lodge.—The farm known as Elvedon Lodge is situated near the road from Brandon to Elvedon, about 3 miles from the former. A large pit is open close by in the Boulder Clay, which clay is used as a top-dressing for the sandy land. In the clay large flints are found, and it is very significant of the local origin of that deposit, that the flints are not cracked or weathered, and are suitable for knapping. Southwell found toppings, upper-crusts, wall-stone and floor-stone, and used many of them. Had these stones travelled far, or been much exposed, they must have become weathered and unfit for knapping; but although striated they are as sound as ever, and the coats are not much reduced in thickness, but the softer, thicker bottom coat has suffered most.*

Thetford, &c.—Flint has been dug near Thetford and gun-flints used to be made in the town, but for 40 years nothing has been done. I have not yet succeeded in obtaining details of the section. Stone was formerly raised, and gun-flints made, at Cavenham and Tuddenham, a few miles S.E. of Mildenhall.

The above include all the stations from which flint is obtained for knapping at Brandon and Icklingham. At Norwich flint is sometimes brought from other pits to Mr. Frewer, when the stone happens to be very good, but as this is only chance trade the sections are not described. Norwich flint has recently been brought to Brandon, owing to the falling off of the local supply.

Mr. Wyatt mentions that gun-flints were made at King Manor, Clarendon, near Salisbury, but there was no regular manufacture. The men worked on the sunny side of the road opposite the pit which yielded the flint, and flaked and knapped in the open air on the spot.†

Gun-flints have been made at Grays, Essex, by Brandon knappers.

My colleague, Mr. H. B. Woodward, F.G.S., has supplied me with an account of the gun-flint manufacture formerly carried on at Beer Head, Devon, from information given by Mr. P. O. Hutchinson. The locality was the undercliff on the west side of Beer Head. "This undercliff," says Mr. Hutchinson, "was one " of the places to which the gun-flint makers used to resort in " order to follow their calling, as the landslip had probably dis- " interred plenty of black flints ready for use, or made them more " accessible than before. Heaps of flint chips and splinters " marked the spots where they had laboured. In rambling " through this place I saw a heap of this refuse large enough to " fill a wheelbarrow; further on another that would have filled " two or three wheelbarrows, and elsewhere others of still larger " size. The men would go out of a morning with their knapping " hammers from Beer and Branscombe, and there work till " evening. I never heard how many they would make in a day.

---

* See remarks on this point in " Geology of the Fenland," and Geikie's " Great Ice Age," 2nd edition. See also " The Fenland, Past and Present," p. 519.
† J. Wyatt in " Flint Chips," p. 588.

" The trade diminished after the Battle of Waterloo, and ceased
" on the introduction of the percussion cap." The chips are now
all dug into the ground, and the locality (which is just below the
words " Signal Staff" on the Ordnance Map) is fast losing all
traces of the old trade.

I understand that gun-flints were made quite recently in
Turkey, but can get no other information about them.

Professor Ramsay tells me gun-flints were formerly made in
Glasgow, but whence the stone was imported he does not know.

Mr. Darbishire, F.G.S., informs me that he purchased strike-a-
lights of the maker in Spain a few years since. They were similar
to the old French gun-flint.

Tools.

The stone-diggers use four tools, a one-sided pick, represented
in Fig. 3, a heavy iron hammer, a shovel, and a short crow-bar,

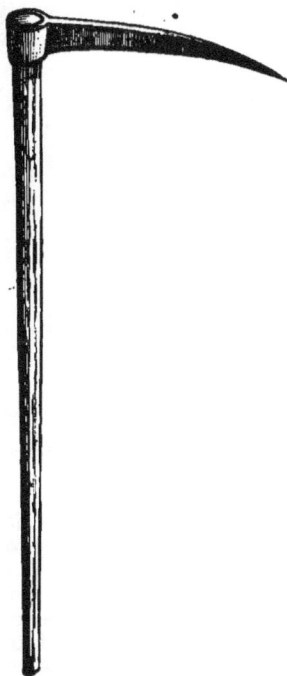

Fig. 3.—*Stone-Digger's Pick.*

none of which calls for special mention; though the pick, as will
be shown hereafter, is interesting from an archæological point of
view.

B 2

The gun-flint makers' tools require more detailed description, and will be described in the order in which they are used, viz., those used (1) in quartering, (2) in flaking, (3) in knapping.

Quartering Hammers are of two sizes, the larger being called the *First* and the smaller the *Second* quartering hammer. The former is represented in Fig. 4.* The weights vary, but those in the collection are average specimens, and weigh 6 lb. 14 oz., and 3 lb. 4 oz. respectively. The hammers are hexagonal in section, and taper but slightly so as to leave the face large. They are made of iron, steel-faced, and when the face wears they are re-steeled. The old, first quartering hammer in the collection shows one used-up face. It has been in use for 20 years. The old, second quartering hammer is in excellent working order, and is about 15 years old. The workmen prefer them in

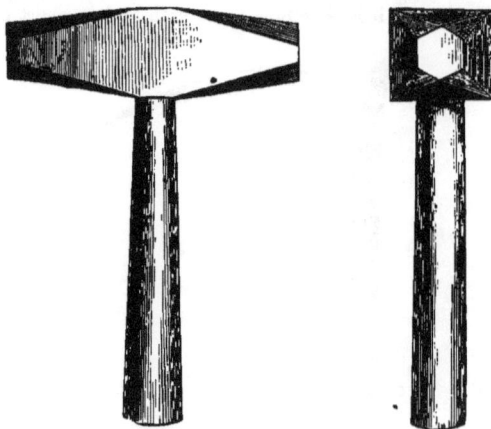

Fig. 4.—*First Quartering Hammer.*

this state to new tools, because the face is worn to the proper shape.

Flaking Hammers are also of two sizes, known as the First and Second flaking hammers. Those in the collection weigh respectively 1 lb. 12 oz. and 1 lb. 4 oz. They are made of steel, have a square section, and taper so as to terminate in a small square face, as shown in Fig. 5. On the centres of the sides the hammers are flattened for the purposes of striking off irregular projections on the quarters. As the blows are given with only a portion of the face that portion wears down until it becomes useless; the other face of the hammer is then used, and when this is worn the handle is taken out and fitted the other way into

---

* All the figures of tools are drawn to ¼ scale.

the eye or socket so as to bring the unused portions of the faces into play. When these become worn the faces are filed up square.

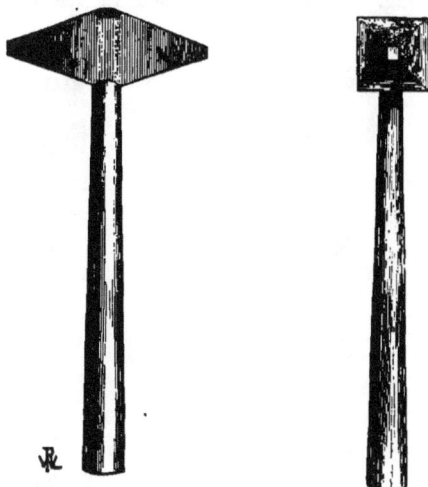

Fig. 5.—*Flaking Hammer.*

These processes are well illustrated by the old flaking hammers in the collection. The first flaker has both faces worked down so as to require filing afresh. It is 20 years old, but has only been in use eight years. It also shows how the sides wear with striking the irregular pieces off. One end of the old Second flaker is worn in the usual manner but has broken from the steel being too hard ; the other face has been fresh filed. This hammer is five years old. With constant working it takes two hours to wear down one side of a face. When the faces get very much worn the hammer is drawn out. The flaking hammers used to be called French hammers, because they were introduced from France. The date of introduction is uncertain, but an old knapper told me he remembered his grandfather saying that a prisoner of war who lived at Brandon, and whom my informant called *Pēro,* was a flint knapper and had some flaking hammers made different in pattern from those used by the Brandon men. These were in all probability the flakers in question, and the time was probably during Marlborough's wars.

With a flaking hammer of given sized face and weight there are a maximum and minimum thickness which cannot be exceeded by any flake struck thereby ; there is likewise a minimum force to be applied to dislodge a flake, a blow of less weight failing to do more than bruise the stone : and a maximum, which if exceeded shatters without flaking the flint. With heavy hammers longer flakes can be struck than with light ones.

English Hammer.—Prior to the introduction of the French hammer the Brandon people used an oval hammer similar to that shown in Fig. 6, which is included in the new set in the collection.

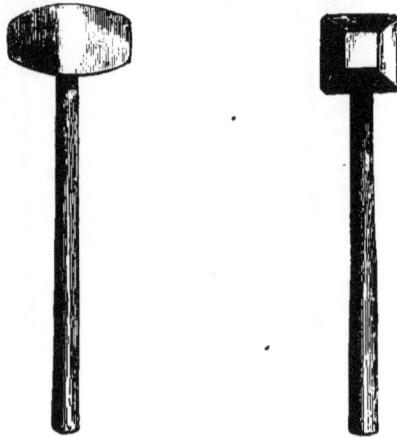

Fig. 6.—English Flaking Hammer.

A hammer called the English hammer is still in use of which the one figured is a specimen, but it is merely a flaking hammer of the French pattern very much worn. The one in question has been in constant use for over 60 years. English hammers are never made specially.

The probability of the old English flaker being a metallic reminiscence of the stone age is discussed in the sequel.

Knapping Hammers are made from 9-inch flat-files drawn out as shown in Fig. 7. They require delicately tempering, or they will fly instantly. Mr. Wyatt states that the heads are set on obliquely. If he refer to the setting of the heads on the handles he has fallen into an error, but if to the set of the hammer on the flint he is correct. The edges at first are quite square, but they soon wear and require filing up daily. They generally become hollowed out in the centre, as is the case with the old one in the collection; much, however, depends upon the individual pecu-liarity of the knapper, each of whom can tell his own by the manner in which it wears. If the edge is not kept square it is apt to gap or split the flints. The old one in the collection has been drawn out, and is very much worn; so much, indeed, as to be fit only for a learner. A quick knapper wears out a hammer in a fortnight. The one in question was originally of the same length

as the new one, and has been twice worn down and once drawn
out. It has been in use for a month.

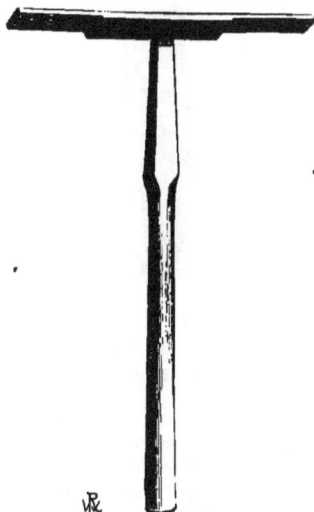

Fig. 7.—*Knapping Hammer.*

The **Blocks** are made of the boles of large elm trees, and
measure about 4 feet in diameter and 2 feet in height.   A small
model block with the stake, &c. full sized is in the collection, and
is represented in Fig. 8.   The blocks are placed by preference

Fig. 8.—*Block and Stake.*

against the wall as shown in Figs. 13 and 18, and slope gently
forwards.   At a distance of about 4 inches from the side the *stake*
is placed.

The *stake* is a piece of iron about 6 inches in length and 1 inch
square at the shoulder, tapering to a point below. Fig. 9 represents
a stake in which *a* is the neck upon which the flint is made, *b* the

shoulder, c the body. To set the stake a round hole is bored 6 inches in depth, and of less diameter than the width of the stake. The

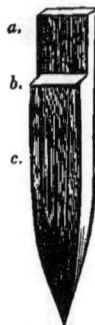

Fig. 9.—*Stake.*

a. Neck.   b. Shoulder.   c. Body.

stake is then made red hot and driven so that the shoulder is from a quarter to three-quarters of an inch above the block. It is then withdrawn, and four pieces of leather, called *stake-leather*, are cut 1 inch wide, 3 inches long, and tapering to a point. The points of these are then inserted into the hole, the stake re-placed and driven home. It must not, however, reach the bottom of the hole or the flakes would strike " dead " on it, and would not work.

Some knappers insert a wedge-shaped piece of oak, about 6 inches by 3, in front of the stake to fix the stage to; this saves the block, which would otherwise be injured by the nails driven into it to secure the stage; the wedge can, of course, be replaced.

The *stage* consists of two parts, the *staging-wood* and the *knapping-leather.* The former is a semi-cylindrical piece of ash about 3 inches by three-quarters, and is nailed on to the block close to the stake so that its length is at right angles to the face of the stake.

Over the staging-wood the knapping leather is fixed. It con- sists of a piece of sole-leather about 4 inches by 1, fixed trans- versely across the wood, by from two to four *wringings*, which are the points of horse-shoe nails wrung off by the blacksmiths. These are always used, because they are rough, have no heads, drive well, and hold tightly, and when the leather is worn up they draw out cleanly, and can be used over again. The width of the leather is such that the knapping hammer falls clear of the nails.

The leather now used is new, and the inside is placed upper- most, because it is rougher and free from grease. Until lately old leather was used. The whole of the stage must be very firmly fixed.

The height of the stake varies with the kind of gun-flints to be made upon it, this height being called the *fall.* For ordinary flints the fall is about $1\frac{3}{4}$ inch; for muskets, 2 inches; but some knappers like a greater fall than others.

In working the fore corner (that is the corner furthest from the knapper, the other being the hind corner) wears down, as shown in the model block, and the hind corner requires rasping down at least once a day to keep it level.

The other implements are the knapper's *knee-piece* described in the account of flaking, and the *flaking* and *knapping candle-sticks*. The former, represented in Fig. 10, consists of a rod of quarter-inch iron, about 5 feet long, with a sliding bracket carrying a candle-holder on an arm with an elbow joint. It is driven into the ground on the left-hand and close to the left foot of the flaker when he is working at night, so that the light shines on to the quarter.

The latter is simply a fragment of perforated tile from a malt-kiln floor. Specimens of both are shown in the collection.

METHOD OF DIGGING.

Division of labour finds no adherents among flint-diggers, each of whom sinks his own pit and raises his own stone; the only aid being the occasional employment of one or two boys, generally the children of the workmen. Two reasons are assigned for this individuality of effort, namely, that the demand is not great enough or sufficiently regular to pay for the use of expensive plant; and secondly, there is much difference in the paying value of the pits, some yielding four or five jags of stone per week while others only yield two or three; hence there is some degree of speculation in the work, and each man hopes to pitch upon a valuable take. When a man is about to sink a pit he takes into consideration the chance of obtaining good stone in plenty, the depth to which he will have to sink, the dryness and warmth of the situation, and the proximity of new or old workings. Of these, the depth is considered perhaps the least, as the quality and quantity of the stone is the most, important. Some men are particular in choosing a take among trees, because the chalk is then drier, and the shelter of the wood diminishes the chance of the workman taking cold when coming up heated to the surface. The proximity of old workings is avoided from the uncertainty of the extent of their burrows, which might seriously diminish the value of a pit, but when very good stone is believed to have been obtained new pits

Fig. 10.—*Flaking Candlestick.*

are sometimes sunk among old ones. Pits are often sunk near together, especially in summer time, when the air is sometimes bad, and the workings are made to communicate, so that a draught is obtained through the two shafts. The extent of the workings is determined by the labour required to carry out the stone got in a day ; they seldom run more than 12 yards in one direction.*

The average time taken to sink a pit of about 30 feet is three weeks, or 10 feet per week, and some of the more careful men commence a new shaft before an old pit is worked out in order to have some money coming in all the time. The pits are worked all the year round.

A digger selects a spot to sink a pit upon, and sets four pieces of chalk, or digs up four sods, at the corners as his "marks," which marks are held sacred and may remain for years before the pit is sunk. No digger may have more than one set of marks besides his pit ; but he may have two pits and a set of marks at one time, and if he clears out the first stage of a new pit that counts as one, though it may not be sunk further for months or even years. Thus a digger can have a pit at work, one begun, and a set of marks.

There is but one other rule observed among the diggers, and that is that none may burrow more than half way towards the nearest pit. Although that man is cursed who moves his neighbours' marks, instances have occurred in which a digger has braved the curse when he has suspected his neighbour of having " spotted a rich pitch," but this is very rare. I have known, on one occasion, a man to begin to sink a pit close to another's marks. Then the enterprise of the diggers shone right royally. The one whose marks were in danger at once commenced his new pit. It was a race for the floor-stone ; for whoever reached it first would at once burrow under the other man's shaft, and the ground would become his. They struck the stone within two hours of each other, after three weeks' incessant work ; but alas, the intruder was to windward of the other, drove a burrow under him, he came toppling through, and the day was won. An amicable fight settled the difference and the original digger moved on.

When a spot is selected, permission is obtained from the trustees of the heath, which is the property of the poor, and the digger commences his shaft. He pays no royalty or "groundage," this falling to the lot of the buyers, who pay on each jag of flint or chalk, and if he fills up his pit after working it out he receives a shilling from the trustees.

The shaft is begun by digging a trench three yards long, one yard wide, and one yard deep.† The long sides generally

---

* Mr. Wyatt says, "The digger tunnels a certain distance (according to the limits of his 'take' or lot), and when he has exhausted the flint-bed of one stage, he works down to the next of the series." Flint Chips, p. 581. In this he has been misinformed, for there are no limits set to the extent of the burrows, and the floor-stone is always worked first, and the others successively as the pit is filled up.

† Mr. Wyatt says 5 feet deep, but the first is always 3 feet and the others 5 feet, except the last, which is more. Flint Chips, p. 581.

Fig 11

Plan of Pits for working out Floor-stone.

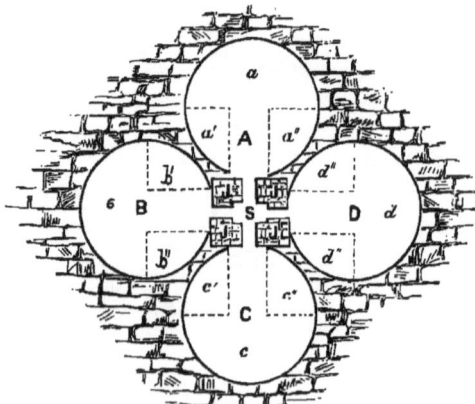

Fig. 12

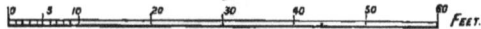

S.B.J.S. *del.*

Plan of Burrowing for Wall-stone.

run *N.* and *S.*, so that the last stage, or " the Two," shall face the mid-day sun. The narrow end of the pit to the left of the Two is called the head of the pit. In the centre of this an opening is made and carried down to a depth of five feet, slightly inclining towards the eastern side of the original trench; a stage is left at this point on the long north side of the original trench. The shaft is carried down another five feet, and a staging left on the short west side; at the next five feet the stage is again on the north side; at the next five feet on the western short side, and so on to the floor-stone, the front and right sides having no stagings. The shaft is only sufficiently large to admit a man, and it inclines or " is on the sosh," so as to undercut about two yards in 30 feet.

If we call the two long sides of the original trench *A* and *B*, and the two short ones *C* and *D*, the " sosh-wise " shaft is along *A* and *C*, *D* and *B* have alternate stages in the series *D*, *B*; *D*, *B*, or, as the workmen say, the stages are made " cross-handed."*

The last stage is called the *Two* and is deeper than the others, some being as much as 9 feet. It probably takes its name from the stone being raised on to it in two heaves, the first being on to a cross-timber.

The object of cutting the shaft on the sosh is to prevent any accident from stones falling from the upper stages. When such a catastrophe occurs, the workman leaning back, plants his shoulder against the next stage, and the stone falls clear of him down the shaft.

The floor-stone is pierced to a depth of about 6 inches and then a gallery or "burrow" is carried slantingly under the stone for about a yard when the burrow is commenced in earnest. Fig. 11 shows the method pursued in burrowing for floor-stone. The first *main-burrow A* is entered through an orifice 18 inches high and 2 feet wide, the floor of which slopes downward for about 3 feet, the roof (from which the stone has been removed) being nearly horizontal. The main-burrow is about 2 yards wide and is driven straight for about 9 yards, and the chalk and stone carried to the surface. At the end of this burrow a "*draw*" 1, is made; that is, the workmen lying on his elbow picks away the flint from above as far as he can reach, thus forming a semi-circular space about 18 inches high; this he continues, and, if the stone be good, he will draw 3 yards in each direction. The stone and chalk from this first draw are carried to the surface. The chalk is always thrown to the head of the pit, and the stone to the foot.

A *side-burrow*, *a*, is then commenced from near the beginning of the main-burrow, and of the same dimensions. It is carried in a curvilinear direction so as to catch the end of the first main-burrow. The chalk and stone are carried to the surface.

---

* This is the normal mode of sinking, but the Two sometimes faces other points than the west. Some men under-cut as much as 6 yards. A stone-digger to whom I pointed out such a pit explained that the man " never sank stunt, but under-ran his two by bubber-hutching on the sosh!" The verb active " to bubber-hutch," signifies to sink more on the slope than usual ; " stunt " means direct, or straight.

About half-way down the side-burrow the first *drawing-burrow*, a, is then made of the same dimensions as at others, and the spaces 2 and 3 are drawn into the main-burrow, the chalk and stone being carried out. From the end of the side-burrow, a, the space 4 is then drawn similarly to 1, but not to so great an extent.

The second side-burrow, a', is then made, and the second drawing burrow, a', and the spaces 5, 6 and 7, are drawn as above, the chalk and stone being carried out; thus leaving all the space between the two side-burrows empty.

The second main-burrow B is then driven, and all repeated as above, but only the stone and large "chalks" are carried out; the smaller pieces, or "fine muck," being filled into the first main-burrow. This second main-burrow is, as are all the burrows, of the same dimensions as the first, but the time gained in carrying the "fine muck" into the main-burrow instead of to the surface enables the workmen to drive the second main 10 yards. When the side and drawing-burrows are completed the space 12 is drawn from the second side-burrow a and the "fine muck" filled into a. The space 13 is next drawn from side-burrow b and cleared out as far as possible. Pillars are shown in the figure at the intersection of the side-burrows; these are, however, not often left, but the spaces J, J, J, J, are always left as pillars or *jarms* to support the roofs.

The above process is repeated in all respects as shown in the figure, the "fine muck" from main-burrow C. and its adjuncts being filled into B, which being larger than A, enables C to be driven about 11 yards. The material from D is filled into C, which being larger than C, enables it to be driven 12 yards. Thus the first main burrow is 9 yards long, the second 10 yards, the third 11 yards, and the fourth 12 yards; the workmen express this by saying they "gain" about three yards in working round a pit.

This somewhat elaborate process is only pursued in burrowing for floor-stone, a simpler plan being adopted in burrowing for the less valuable wall-stone or toppings.

When the floor-stone is exhausted, the pit is generally filled in to the level of the wall-stone, which in consequence of its legs, is burrowed from above. A main-burrow A, Fig. 2, is driven for about three yards and the stone and chalk carried out. The space a is then drawn, and the drawing continued right round as shown in a, a', and a'', the material being carried out.

The second main-burrow B is then made of the same size and length, and the chalk filled into A; C and D are then made in the same manner. Nothing is "gained" in this mode of working, the object being to clear out as much stone as possible in the least possible time. The jarms J J J J are left as before.

The pit is now filled up to the toppings, unless building-stones are in great request when the upper-crust is burrowed. The same method is adopted as in the case of wall-stone, but sometimes the stone is merely drawn round as far as the workman can reach. After the toppings are got the pit is filled in, nearly to the surface, the workman receiving 1s. for this job. Toppings and upper crusts are worked from below.

The diggers use a small, one-sided, steel-tipped iron pick, with which they cut the chalk and clear it away from the stone, which is then prised down with a short thick iron crow. The flint generally comes away in pieces too massive to carry out, and these are broken into suitable sizes by an iron square-headed hammer weighing from 5 to 7 lbs. A flint pick is represented at Fig. 3. In the burrows the men sit and pick by the light of a candle, but in the draws they have to lie sideways resting on their left arm and working with the right. Happy is the thinnest man for he has less dead chalk to pick for the same quantity of flint.

Each day's proceeds are carried to the surface, and it is generally enough to fill five stages, three with chalk and two with flint. The pieces are carried up on the head, the chalk first. A lump is carried three stages up and deposited, another is fetched, and so on till the stage is full, then the lumps are carried three stages higher, and then out of the pit.

The chalk is thrown into a heap at the head of the pit, and used for filling in the pit. It is seldom bought, as it can be obtained from pits in the town. Sometimes, however, the squarer pieces are sold for rough walling, and a groundage of 4d. is then paid to the trustees by the purchaser for each "jag" or one-horse cartload.

The flint is brought up in pieces averaging 2 feet by 18 inches, and stacked edgeways on the ground round the pit's mouth. The stacks are covered with dried bracken and fir-boughs to prevent the sun and wind getting access to the stone, and cracking them or turning them milky in colour, for only black flint is used for the best gun-flints, though the milky-coloured stone is equally good. The merchants will only buy black flints.

A good pit lasts from six to nine months. The burrows are never timbered, and accidents from falling roofs are rare.

The stone is sold by the diggers to the flint knapper's by the jag, which, as before stated, is a one-horse cart-load about equal to a ton. The following are the present prices (January 1876):—

Floor Stone:

				s.	d.
Stone	-	-	-	6	6
Groundage	-	-	-	1	8
Cartage	-	-	-	1	0
				9	2

Wall Stone:

				s.	d.
Stone	-	-	-	5	0
Groundage	-	-	-	0	10
Cartage	-	-	-	1	0
				6	10

Toppings:

				s.	d.
Stone	-	-		3	6
Groundage	-	-	-	0	10
Cartage	-	-	-	1	0
				5	4

Upper Crust:

				s.	d.
Stone	-		-	3	6
Groundage	-			0	10
Cartage	-	-	-	1	0
				5	4

The cartage is irrespective of distance; upper-crust flints are only used for building-stone. A jag of stone of ordinary quality will make about 6,000 gun-flints, one of good quality 12,000, and a jag of very good stone has been known to yield as many as 18,000, but in these small-sized gun-flints were included. The bulk of the waste is about equal to that of the raw material.

A jag of stone selected at random by me, but of fair average quality, weighed 13 cwt. 1 lb. It yielded as follows:—

----	Number.	Weight. cwt. st. lbs.		
Flakes -	8,800	—		
Gun-flints	10,850	1	6	0
including				
Second Muskets	550	—		
Common black Muskets -	500	—		
Grey do	150	.—		
Second Carbines - -	4,100	—		
Common do - -	1,900	—		
Horse Pistols - -	2,000	—		
Singles -	400	—		
Large Guns	750	—.		
Small do. -	500	—		
Faced Builders	60	0	6	13
Rough do. - - -	210	3	3	4
Chips	—	6	7	12
		13	0	1

The larger waste,* consisting of outside and bad flakes, irregular pieces, &c. is used for road-metal, and fetches 6d. a jag. The finer chips from the gun-flints are used for garden paths, &c., and until lately were given away, but now fetch 4d. per jag.†

---

* The waste amounted to 53 per cent. of the whole, and the gun-flints only to 13 per cent.

† Immense heaps of waste are still to be seen in Brandon, although thousands of tons have been used on the Great Eastern Railway, and every available pit is filled up with it. Some years since hundreds of tons were carted on to Lingheath; the presence of masses of chips does not necessarily show that gun-flints have been manufactured on the spot.

The flint is very variable in quality and in constancy of colour. Some is quite black *in situ*, but changes to a milky tint in the course of an hour, while other stones never lose their black colour even when exposed to the weather. Of these some of the best examples are to be seen in the north wall of Lode Cottage, which I occupied at Brandon, and which are as black as when dug 20 years ago. These were obtained from very deep pits in the Poors Plantation, in which the flint was larger, blacker, and better in many respects than has been since obtained. There is little doubt that if this house were pulled down, the flints in question would be worked up into gun-flints. This stone had good running qualities, and flakes 8 inches long were not uncommonly obtained from it. A few outside flakes from it are shown in the collection.

MANUFACTURE.

In the process of manufacture there are four processes—*drying, quartering, flaking*, and *knapping*.

Drying.—The stone is brought from the pits and shot down outside the shops where in the summer it dries very rapidly, but in the winter remains wet. In the summer the stone is brought inside the shops and often worked at once, but it is now becoming the practice to sprinkle water over the blocks; the reason assigned being that it "lays the dust." The quarry-water, however, is always got rid of, and hence in winter the stone is stacked round the fire-place to dry. Water is not sprinkled on the stone in winter, the flint being so hygroscopic that it is always damp outside. Even the dry flakes will turn damp when the atmosphere is moist, and then they are sometimes dried over again.

Quartering.—A block of stone is then taken and quartered. The workman sits on a stool placed in front of a window or door, and is very careful about setting it slightly sloping forward so as to incline the body, and obviate the necessity for bending, which would lead to back-ache. Some workmen are very particular about their stools, and will spend half-an-hour in the adjustment; and in summer, when the stool is placed opposite the door, will rather jump over it than shift it.

The workman wears a large leather apron, and on his left knee a *knee-piece* made of pieces of old boot-tops about 6 inches by 12, on the top of which is a *cross-piece*, made by preference from a stout boot-sole, or failing that from a piece of old harness. The cross-piece is about 6 inches by 3 inches, with a hole at each end by which the entire knee-piece is bound lightly across the knee with a leather thong called a *knee-piece string*.

The blocks of stone are taken just as they are delivered, and vary in weight from about a quarter of a hundredweight to nearly two hundredweights. One is placed against the knee so as to bring a flat or hollow part upon the knee-piece for the hammer to strike upon. It is then slightly tapped with the quartering hammer, the large or small one being used according to the size of the stone. The tap tells the workmen whether the stone is sound or not. If it is full of cracks it flies to pieces with a jarring sound. If the coat is hard and the hammer rings, the stone is sound. If the hammer falls dull and jumps, the stone is sure to be double-coated, and grey or mixed colour beneath the coat.

Sometimes pieces fall out on tapping the stone which have a sub-conical inner surface; these are known as *pot-lids*.

The stone is then quartered or broken into pieces of a convenient size to work. The blow is given from the elbow, the hammer being raised about a foot and allowed to fall, little or no power being put into the blow. The stone is nearly always struck from the natural upper surface, because the bottom coat is softer, and the hammer does not *bite*. If, however, as occasionally happens, the bottom coat is the harder, it is quartered from below.* The stone has to be broken so as to leave a more or less square edge to begin flaking from. The quartered pieces average about six inches square, but there is no regularity in the size.

Flaking.—The next process is flaking, which is performed on the same stool as was the previous process, the workman quartering a stone and then flaking it up. This is the most difficult branch of the business, and requires great skill and nicety of judgment. The stone must be struck at the proper angle, in the exact spot, with a certain force, and by a given portion of the face: and all but the first of these elements vary with every flake. Many knappers are unable to flake, and but few attain great proficiency in the art. The quarter is grasped in the hand and the face brought against the knee-piece at an angle of about 45°. The blow is then given by raising the flaking hammer (large or small according to the size of the stone) from the elbow about 2 inches, and allowing it to fall by its own weight, or with a slight extra force according to the size of the quarter. The stone is struck squarely, but not with the whole face of the hammer. If the flake is to be thin, the blow falls just inside the face; if thicker, a little further in, but a flake could not be struck to any purpose if the whole face fell on the stone, and this is the limit of thickness for flakes produced by any hammer. The outside flakes, called "shives," which show the coat, are thrown aside as waste, and by removing these the block is made to assume a rough, many-sided, polygonal form. The next series of flakes are so struck as to fall a little to one side of the previous blows, and the flake runs so as to include the angles or ribs of two of the first flakes, and it is thus double backed. The edge of this flake leaves another rib.

---

* The best stone has a thin hard upper coat and a thick soft bottom coat. The coat may be quite white or brownish below and is sometimes bluish, but these latter are not good and are generally "shotten bottoms" or pitted as if by shot. The tops of good stone are often bluish.

Fig. 14.—*Front View of Core, with Flakes replaced, showing the Points of Percussion.*

Fig. 15.—*Side View of Flint Core (Fig.* 14).

The next flake is struck a little to one side of the similar one in the previous series, and so on. In this way the flaker works round the quarter and removes from two to three rows of flakes, according to the quality of the stone. Figs. 14 and 15 show a *core* with the flakes, 38 in number, replaced, and the points of percussion are plainly traceable and show how each blow was struck. Simple as this process appears, it is, nevertheless, very difficult; for great precision of judgment is required, not only to determine the nature of the blow, but so to flake that the greatest number of useful flakes can be obtained from a single quarter. The stone varies in quality, some running well and clean, others breaking off short and "stubbly;" and unless the flakes are struck of different sizes much waste would ensue. It is this judgment which distinguishes a good from an inferior flaker; a good one would work to profit stone upon which an inferior man would lose money. The cores resulting from the flaking are squared or rounded up, and used as building flints.

The workman sits with five small tubs around him, two on his left, and three on his right. Into the hindermost left-hand tub, called the *chip-tub*, the waste irregular fragments are dropped, this tub being close at hand under the quarter so that the waste will fall into it; the larger pieces are thrown into the other tub, called the *builder-tub*, which is at arm's length from the flaker. The three tubs on the right are placed in a triangle close together; the left-hand one, called the *best-tub*, receives the good double-backed flakes, the middle one, called the *common-tub*, the single-backed flakes (of which many must be struck), and the right-hand one, called the *little-'un-tub*, the small flakes which are given to apprentice boys to practice upon. (*See* Fig. 13. Frontispiece.)

Considerable skill is also required to reduce the amount of work to a minimum; the hammer is made to fall on the near side of the knee-piece; it is then dropped on to the leg, while the flake is thrown into the tub, and lies with its face over the spot on which it is first tapped, so that it is ready for the next blow; in this way much time is saved. The flattened side of the flaking-hammer is used to dress off irregular pieces.

The flaker works according to his needs. If he has to supply a quantity of flints of one size he makes his flakes accordingly. Thus he will flake for muskets, carbines, horse-pistols, or even singles.* As a rule, however, several kinds of flints are required together, in which case he makes large and small flakes. Single-backed flakes are never struck intentionally. The first two flakes must necessarily have only one back; but if the stone be good no others are made from a quarter. A flake to be perfect should have the face flat, the edges even, and the ribs parallel all the way down, but this can only be obtained from very good stone. As a rule the ribs are not parallel, nor the edges quite straight, and many stones will not run with a level face, but are twisted or

---

* It would be impossible, however, to strike all the flakes of a given size without wasting a great deal of stone.

curved, when they are said to be *wrung*. When a flake does not run, but breaks short or turns at the end it is said to *dub*.*

Fig. 16.—*Double-backed Flake,*   Fig. 17.—*Single-backed Flake,*
         natural size.   *natural size.*

These two flakes have run right through the core, and are hence square ended.

When the quarter is worked up an edge is left; the stone is then turned over, and a flake struck off with the English hammer. Such flakes are called English flakes and have no ribs, and are used for making common flints.

An average flaker will make from 7,000 to 8,000 flakes in a day, a good one 10,000, or working long hours, say from 6 a.m. to 10 p.m., 12,000. Dorling and Southwell have made 60,000 flakes each in one week, and the former has made 63,000 in the same time.

A good flaker works so fast that by the time a flake falls into

---

* Figs. 16 and 17 represent very beautiful flakes which depart from the ideal type but little.

To face page 31.

Fig. 18.—*Knappers at Work.*
(From a Photograph.)

the tub a fresh one is struck off, as they say " the sound of the hammer and of the falling flake should be heard together;" very few, however, can attain to this degree of excellence, and very good stone must be used. Flaking is the most difficult part of the manufacture. The modern flakers excel the Neoliths in skill, as is shown by a comparison of their respective residual cores. This is not the result of improved tools merely, for I have seen Southwell flake as accurately with a round pebble as with a steel hammer.

Knapping.—The final process is *knapping*, or the forming of flakes into gun-flints. The knapper sits on a stool at the block, and is equally particular with the flaker as to its position. The block slopes gently towards the flaker, and the stake slightly inclines in the same direction. Until recently the stake was set upright, but it is now becoming the practice to incline it. The flaker sits close to the block and at right angles to it. The left leg is extended parallel and close to the block, and the right leg is bent. The workman wears a large cotton apron over both knees, and hitches one end of it on to a tack behind him; this apron catches any pieces which fly. The flake is taken in the left hand, and the knapper tells at a glance and by the touch how many and what sorts of gun-flints the flake will make. A good flake will make four, and a very good one five flints (see Fig. 19); 32 were made by Southwell out of eight flakes running. Occasionally four carbines are made from one flake, and Southwell has made four carbines and a horse pistol from a single flake recently. The French flint-knappers are said only to make one gun-flint from each flake.*

The flake is held on the stake face uppermost, the inclination varying with the amount of "running under" required. If the flake is held at right angles to the stake it cuts square, if inclined it runs under, and the greater the inclination, the greater the running under.

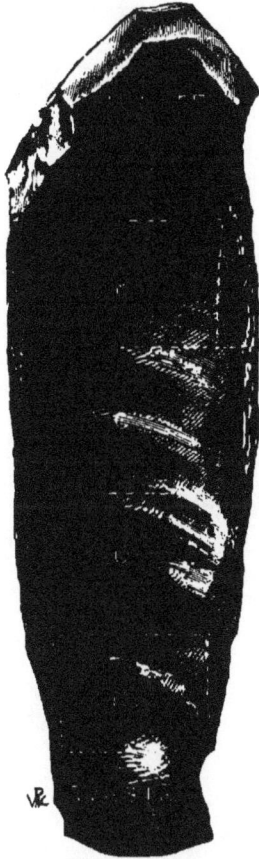

Fig. 19.—*Flake marked to show how it is to be knapped.*

* This need not imply inferior workmanship for their tools, and consequently their flakes, are smaller than ours. As a matter of fact, however, they are less well made and millions have been sent to England to be re-worked.

The elbow of the knapper rests in the groin, the fore-arm is kept perfectly steady, and the motion of the knapping-hammer is entirely from the wrist. The part of the hammer nearest the body is called the *hind corner*, the opposite end the *fore corner*. The flake is struck squarely on the face just inside the stake, so that a shearing force is applied, and the hammer lies with the hind corner and about half the edge on the flake. If the flake is thick the fore corner of the hammer is used, but not otherwise. The flake is first tapped or *tiddled* to get the hammer square, and is then cut at a single blow. Mr. Evans supposed the tiddling was to slightly notch the back,* but this is an error, as the back is never so notched. The blow cuts the flake, but in consequence of the shearing force, the bulb of percussion is formed on the back by the stake, and not, as might be expected, by the hammer. The motion of the hand is kept up continuously, and while the knapper is picking up a fresh flake he keeps on tapping on the leather stage. The sound of knapping is very peculiar, the strokes on the flint yielding a clear musical note, those on the stage a dull thud, and as these sounds never cease and are very peculiar, they cannot be mistaken for any others. When the flake is twisted or *wrung* the body and wrist of the flaker are inclined so as to bring the hammer face square.

The workman first determines which side of the flake is to form the edge, choosing the straightest and best for that purpose, and holding the flake so that the edge is away from him, the face of the flake being uppermost. The edge and heel are made from the sides of the flake, the sides by the cut section of the flake. The first operation is to cut the flake across to form the right side of the flint (the flint being looked at with the heel nearest the body) and chisel† it straight with one or two slight blows. The flint is then turned towards the knapper and the heel cut and chiselled, this being done so rapidly that it is almost finished during the act of turning. The flake is again turned towards the body and the other side cut. Turning again in the same direction and at the same time turning the flint back uppermost the edge is trimmed. These motions are so quick that the flint never seems to stop in being turned, yet every blow has to be considered and delivered according to the individual necessity of the case. The edge is often put on by scraping the face along the back of the stake, and though in many instances this trimming is unnecessary the knapper almost always does it from custom. So rapid is this

---

* Ancient Stone Implements, p. 19. It is probable he was misled by being told that the stake made "*knots*," *i.e.* bulbs, on the flake, and mistaking the word for notch.

† The motion is in the opposite direction to that of the hands of a watch. The *cutting* blow cuts the flake clean across, the bulbs being made by the stake; the *chiselling* is done by a series of light taps which break off tiny chips. Mr. Evans's description (Anc. Stone Imp., p. 19) is incorrect. The heels of French gun-flints are chiselled and not cut. If a flake be good the edge and heel are ready made, and only require chiselling straight, but much depends upon the kind of gun-flint, single-edged ones nearly always require the heel cutting. The edge, being formed by the side of the flake, is never cut.

process that I have on several occasions timed Southwell and seen him make eleven well-finished gun-flints in a minute, and once he completed thirteen. This extreme speed would not, however, be kept up continuously, and eight per minute may be taken as the average for a good workman like Southwell. Inferior knappers sometimes cut the edges and sides, and then go round them again to chisel or trim them.

Each knapper has a certain peculiar style which enables him to distinguish his own flints. He tells them at once by feeling them with his left hand, and though the differences are very slight I have always found a knapper correct, though he cannot say more than that a gun-flint is or is not his own workmanship. This judgment is irrespective of sight and can be equally well used in the dark. I tested Southwell blind-folded with about 30 flints made by himself, myself, and three other knappers, and put with them three which I had made but he had finished up for me. He was right in every case. When he came to one which was a joint production he hesitated, and said "I seem to have handled this, but there are points about it that don't seem to be my work." This struck me as being very extraordinary. The flints were given to him one by one, and he did not see any of them, nor was he told whether his statements were correct until he had finished.

It will be noticed in carefully examining flints that one side is nearly always less well made than the other. This is because flint does not cut so well after it has been handled.

An average workman will knap 3,000 flints in a day of 12 hours, but a good one will make 4,000 at a pinch. One man for instance, working from 4 a.m., till 11 p.m. made 24,000 in a week. Southwell and a boy, with the aid of an extra man and a boy at night, made 44,800 in a week, flaking included.

Fig. 18 shows Mr. Southwell in the act of knapping and well illustrates the position, &c. The flints when made are thrown into tins ranged round the block as shown in the plate. Commencing on the left of the knapper the tins are so placed that best and most used flints are placed in the tin opposite to him. Mr. Southwell's arrangement is 1 Second Single, 2 Second Carbine, 3 Second Horse-pistol, 4 Second Musket, 5 Common Carbine, 6 Common Musket, 7 Common Horse-pistol, 8 Common Single, 9 Grey Musket.

When the tins are full the flints are counted. This is done in fives called *casts*—an average tin holding about 100 casts. The flints are thrown on to a board or table, at which the counter sits and draws the flints into his apron, three with the right hand two with the left hand. The process is very rapid. Southwell counted in my presence 700 in 1 minute 35 seconds, or at the rate of about 1,000 in two minutes. He considers this about his average speed and has counted 50,000 at the same rate, including the time spent in picking up, emptying, and replacing the tubs or cans.*

---

* The quickest counter in Brandon can count 600 a minute.

At one time a leather collar or disc was placed on the knapping hammer close to the head to prevent chips hitting the hand, but this is never done at present except by old men.

The prices paid to journeymen are, per thousand, for flaking, a shilling; for knapping, 14 pence, but in slack times the price has fallen to seven pence. The prices of the gun-flints per thousand at present in the town are :—

				s.	d.
Second Muskets -	-	-	-	3	6
Common Muskets -	-	-	-	2	9
Spotted Muskets -	-	-	-	2	6
Solid Grey Muskets	-	-	-	2	0
Second Carbine -	-	-	-	3	4
Second Horse-pistol	-	-	-	2	9
Common Horse-pistol	-	-	-	1	8
Second Single	-	-	-	2	6
Second Double -	-	-	-	1	6

The gun-flints are packed into tubs or sacks containing from 5,000 to 20,000 each, and are consigned to merchants in Liverpool, Manchester, Birmingham, Bristol, and London, by whom they are shipped for foreign markets, very few being used in England, these being chiefly used for horse-pistols. The old gunners or duck-shooters about Brandon still use flint-guns.

BUILDING FLINTS.

The cores from which flakes have been struck are slightly worked up to form building-stones, or *builders*, as they are more commonly called. From their mode of formation they taper towards the end ; but they are worked to a level face if possible, but if any projection exists on the face it is struck off and is then called a " chip-back," because it has been chipped back to a level face. These, however, are not considered so good as those which have a smooth face struck by a single blow of the quartering hammer. Builders are known under the following designations :—

1. *Square Black-faced Builders.*
2.   „   *Mixed coloured*   „
3. *Round Black*   „
4.   „   *Mixed coloured*   „
5. *Randon faced*   „
6. *Rough Builders.*
7. *Land Stones.*

The first four are made to size as ordered, but this size is arbitrary ; the fifth are called randoms, because they are of any size and not assorted. The two last are not cores. No. 6 are the irregular pieces broken from the quarter, and No. 7 are merely stones picked from the surface of the ground, and built into walls without any dressing.

These flints make a very durable and nice-looking wall. Many of the houses in this district are built of them; and, when the quoining is of brick, the appearance is quite attractive, and possesses none of the heavy look that might be expected. Faced flints have been used for building purposes for a long period, and some of the Norwich churches show them very cleverly worked to fit into freestone tracery of the Decorated style. The trade in builders is steadily increasing, and their durability and effectiveness, combined with small cost, bid fair to give them a more extensive sale than has hitherto been their lot.

The prices in the town at present are:—

			s.	d.	
For	*Square Black-faced Builders*	-	4	0	per 100
„	*Mixed Coloured ditto* -	-	4.	0	„
„	*Round ditto* -	-	4	0	„
„	*Mixed coloured ditto* -	-	4	0	„
„	*Random faced*	-	3	0	„
„	*Rough Builders*	- 3s. 6d. to 5	0	per ton.

Land stones only bear the expense of cartage. In Suffolk, mixed coloured builders are preferred to black ones.

Specimens of each of these are shown in the collection; and, on the square black-faced builder, Southwell has marked his initials and the date of the formation of the collection. This is done by means of sharp re-bounding blows with a punch, and the shape of the indentations shows that each blow drove out a little bulb.* This is an old method of perpetuating the name of the maker of builders. For instance, on Icklingham All Saints Church there are seen—

J. B. (= James Benstead) 1806.
H. A. (= Henry Ashley) 1865.

On walls built of flints small circular pittings, seldom more than a quarter of an inch in diameter, are often seen. They have been attributed to the action of frost upon the stone, but I feel sure that no exposure to mere weather would ever cut out such depressions, but that they are produced by slight blows. Each will be seen to possess a tiny bulb of percussion, and the pittings are always most plentiful within a few feet of the ground, or in just the position in which the stone is most likely to be accidentally struck. Higher up, out of reach of ordinary mishaps, they are much less frequent, although they are more exposed to the weather. It is true that some stones high up in a building show pittings, but building-stones are not carefully preserved from injury, and some of the specimens in the Museum of Practical Geology are smothered with depressions, and these have clearly been tapped on the face by the builder to bring them into line with the others. On a smooth surface the weather could not possibly produce the pittings; a crack might be enlarged by the expansion of frozen rain-water, but no bulb

---

* Gentler blows, which do not detach the pieces, produce incipient bulbs which give to the flint the peculiar appearance known as "snake pattern" in agates. This pattern is produced in the same way. A specimen of flint so marked by me is in the collection.

of percussion could in any way be the result. I have found by experiment that the slightest blow, if of a rebounding character, will produce incipient bulbs, even pebbles the size of a pea projected with moderate force are sufficient for this purpose, and water finding its way into the little fissure so produced would on freezing enlarge the fracture, *but its direction and extent would be determined by the momentum and area of striking surface of the original blow;* hence the piece will assume a conchoidal form such as is invariably found to be the case.

I have long and carefully studied the fracture of flint, and feel certain that bulbs of percussion can only be formed by blows. Around Brandon it is very common to find flints naturally fractured, with one surface studded all over with pittings similar to those in question, and they are most frequent in the vicinity of old flint workshops. A specimen of these is shown in the collection, and it will not be surprising if they turn out to be stones used in detaching the small flakes from the surfaces of arrowheads, &c.

Miscellaneous.

Strike-a-lights are shown of four shapes. Fig. 20 represents

Fig. 20.—*English Strike-a-Light.*

the usual form of the old English kind, which is identical with the present French strike-a-light and the modern French and old

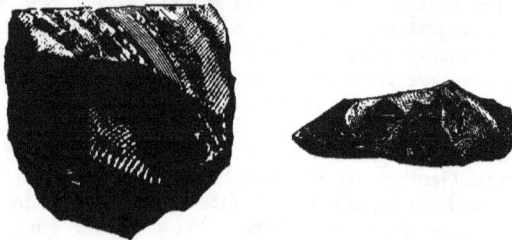

Fig. 21.—*French Strike-a-Light.*
See also Fig. 59.

English gun-flint; and, as will be hereafter shown, very like some of the ancient so-called "scrapers." Figs. 21 and 22 represent particular shapes which are sometimes ordered to this day and are again similar to some old "scrapers." These strike-a lights are made from English flakes, and the first gun-flints only differed from them in being of smaller size; the old gun-flints were, in

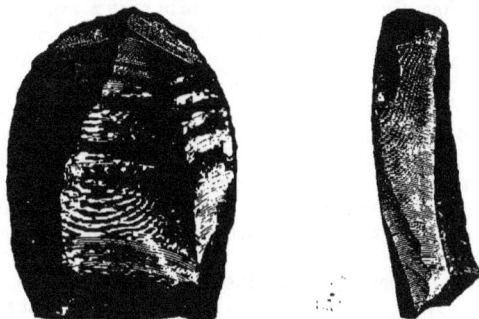

Fig. 22.— *Horse-shoe Strike-a-Light.*

fact, only modifications of the existing strike-a-lights. I venture to name Fig. 20 the English strike-a-light, Fig. 21 the French strike-a-light, and Fig. 22 the Horse-shoe strike-a-light, for obvious reasons, and because it is as well to have names for these peculiar relics of the stone age.

I have a German gun-flint which is also identical in shape with some of the so-called " scrapers."

Specimens of discoidal flints are shown of which many thousands were made some years ago for a London merchant, but their use is unknown. They are supposed to have been used for some ornamental inlaid work, or for pivots on which axles were to work.

The waste chips are used for road-making, and were formerly given away, but are now sold at 3d. per one-horse load. There are thousands of tons still lying around the knappers' shops.

Attempts have been made to introduce these chips into the pottery districts for calcining and using in the manufacture of china, but they were not very successful as so many of them were outside chips retaining the coat, which was not serviceable. I believe another attempt is about to be made to re-introduce them.

At present large quantities of brown weathered flints are imported from Havre for this purpose, and it is thought that the native material can be sorted and utilised at less expense.

NOTES ON FRENCH GUN-FLINTS.

The French knappers (*caillouteurs*) make use of the following tools :—

" 1. A hammer or mace of iron, with a square head, the weight of which did not exceed two pounds (but it may be of half that weight only), with a handle seven or eight inches long. This tool is not made of steel, for an excess of hardness would render the strokes too hard or dry (as the phrase is) and would shatter the nodules irregularly, instead of breaking them by a clean fracture." This implement answers to the present " quartering hammer." It was, however, smaller than that tool, and leads me to conclude the flints were brought to the knapper in smaller pieces than to their English brethren.

2. A hammer with two points. This is made of good steel, well hardened; its weight does not exceed 16 ounces Its handle is seven inches long, passing through it in such a manner that the points of the hammer are nearer the hand of the workman than the centre of gravity of the mass." The present "flaking-hammer" resembles this tool. The points of the flaking hammer, however, are not inclined, and the tool is lighter than the English one

3. The disc-hammer, or roller, a small tool, called in French *roulette,* which represents a solid wheel, or segment of a cylinder, two inches and four lines in diameter : its weight does not exceed 12 ounces. It is made of steel not hardened, and is fixed on a handle six inches in length, which passes through a square hole in the centre.

4. A chisel, tapering and bevilled at both extremities, seven or eight inches long, and two inches wide, made of steel not hardened. This is set on a block of wood, which, at the same time, serves as a bench for the workmen. This chisel answers to the "stake" in present use.*

The only place in which gun-flints are now made in France on a scale of some magnitude is the village of Meunes.† (I have since learned that so late as 1870 gun-flints were made at Romantin (Loire et Cher), Lye (Indre), and St. Aiguan (Loire et Cher). Whether the manufacture still continues I do not know. I have samples of this work, and they are identical with the French gun-flint figured below.)

The stone for manufacturing the gun-flints for the French army was selected with great care. It was obtained from Meunes, Couffy, Pouillé, Angé, Châtillon, Noyers, Langon, Lyes, Paulmey, Lucion, and Valençay. At the first-mentioned place a special artillery officer was stationed to examine the gun-flints. Moyesse and St. Vincent (Ardèche), Cérilly (Yonne) and La Roche Guyon on the banks of the Seine (Oise), are also mentioned as furnishing good beds of flint.‡

The flint-pits on the banks of the Cher were from 40 to 50 feet in depth, from whence levels or horizontal galleries were driven "into the only good stratum which is known in that district," but whether above or below the flint is not stated.§

"In parts of France three or four workmen used to join for the purpose of excavating flint for the gun-flint makers. They would first dig a trench, about six feet in length and depth, and about two feet in breadth; then they would make a second trench lower than the first, and so on, like the parallels of siege-works, till they got to the depth of from 30 to 40 feet, where the flint nodules were found embedded in a soft kind of chalk."‖

The specimens in the collections show that the French gun-flints are nothing but small strike-a-lights.

---

* Rees, article Gun-Flint, op. cit.
† Steenstrup and Lubbock.   Flint Implements of Pressigny, Trans. Eth. Soc., N. S. Lond., vol. v., pp. 221–227
‡ Arch. Dép. Cent. de' l'Artillerie.
§ Rees, article Gun-Flint, op. cit.
‖ Wyatt, Flint Chips, p. 582.

ANTIQUITY AND DEVELOPMENT OF THE FLINT TRADE AT BRANDON.

From palæolithic times to the present day the vicinity of Brandon has been one of the great emporia for flint, and the huge neolithic pits at Grime's Graves attest by their size and number how important the manufacture was in those times. When we remember how largely flint was worked, even after the introduction of metal, it is highly improbable that its fire-producing value should not early be discovered, and this would give a fresh impetus to the flint-trade, and strike-a-lights would become merchantable commodities, and we know they were made here before the introduction of gun-flints.* From this trade the manufacture of gun-flints is a lineal descendant, and it exists at the present day. The forms of strike-a-lights have not varied since neolithic times; the first gun-flints were nothing but small strike-a-lights, and the modern French and German gun-flints are still of this character. This seems to be the true genealogy of the gun-flint, and show the intermediate steps between the old strike-a-light and the modern English gun-flint are very clear.

There is, I believe, no evidence of palæolithic man mining for flint, but the neolithic people (separated in time from the old stone folk by an interval perhaps equalling, if it did not exceed, that which parts us from them) certainly mined largely, and I cannot but think that the manufacture of flint implements for various purposes has gone on without any, or but slight, interruptions from the earliest times to this moment.

It becomes a nice question to determine how far the process has improved or degenerated and though from the nature of the inquiry but little can be determined, that little is of great interest.

So far as mining is concerned the record is one of steady but slow progress. The diggers of the old flint-pits at Grime's Graves sunk funnel-shaped shafts to a depth of 30 feet, with a diameter of 25 feet at the top, and this Herculean task was accomplished by the aid of no better tools than unbored stone celts and picks made from the antlers of the red deer. The sides had no stagings, and the material was in all probability hauled up in wicker or skin baskets by sinew or fibre ropes. They drove simple 3-foot burrows on the top of the floor-stone and did not *draw* the ground, neither did they drive their burrows straight, but crookedly. It appears to me very probable that several adjacent pits were worked at one time, each by its own set of men. The flint was worked on the spot† and bartered to the hunting

---

* Mr. Evans also advances this opinion (Stone Imp. pp. 16 and 283), but although he shows the similarity between modern strike-a-lights and some of the ancient " scrapers," he does not allude to the identity of the former with old English and modern French gun-flints, which is the connecting link between the ancient and modern industries.

† Only the larger implements, such as celts, seem to have been made at Grime's Graves. The smaller flakes were taken down to certain picturesque spots on the river side and there worked into arrow-heads. I have found several of these workshops, and obtained quantities of tiny flakes from the arrow-heads, also fragments of pottery, burned stones which were probably used as pot-boilers, pieces of charcoal from old fire-places, and fragments of nut and bone. I am not aware that this fact has been before noticed.

tribes, and the magnitude of the works is shown by the immense number of flakes and cores which literally strew the ground in the vicinity of the old pits. It was owing to this permanent employment that the art of flint-implement making attained that degree of excellency which culminated in the beautiful ripple-chipping of the surface shown in the annexed cuts, Fig. 23 being

Fig. 23.—*Flint Arrow-head from Chatteris.*

from Chatteris and probably of Brandon flint, and Fig. 24 from

Fig. 24.—*Flint Arrow-head, Brandon.*

the vicinity of Brandon itself, and certainly of native stone. This exquisite specimen was kindly lent to me for engraving by my friend Mr. H. R. Maynard who possesses some very fine examples of neolithic art.

So far as mining is concerned there has been a decided improvement since Grime's Graves were dug, but the progress has been very slow, and, I believe, not sufficient to mask the primitive origin. Thus the independent character of the pits and their close proximity are exactly paralleled in Grime's Graves : the similarity in the appearance of the neolithic pits and the deserted modern ones is very striking. The method of lifting the stone is almost or quite as simple as that of yore. Then it was hauled up by a simple mechanical contrivance, now it is lifted from stage to stage, by manual labour. In the method of getting the stone, however, the progress has been decided as shown by the elaborate method already described ; but even here, although the work has been carried on so long in a faulted rock, the diggers have not discovered the simple law which determines the upthrow and downthrow.

When flint was first used for fire-arms it was merely broken into a convenient size, and each soldier had to find his own; in like manner it is probable that when flint was first used as a material to form weapons conveniently shaped natural stones were selected. When the gun-flint trade became a special branch of industry the workmanship rapidly improved in the hands of workmen who devoted themselves solely to it, and is now brought to a state of perfection never before equalled.* So with the old flint-implement trade; it began to improve rapidly when men devoted themselves entirely to it, and culminated when metal gave them better tools and before it became too common to replace the stone. The beautifully even surface-chipping of such arrow-heads as Figs. 23 and 24 is a triumph of skill and a proof of luxury; and the art is lost to us. No Brandon knapper can in any way approach it, as Mr. Evans has said, and as I have verified. So far there has been a degeneracy in the working of flint.

I cannot help hazarding an opinion which has been formed from an intimate knowledge of flint-knapping and of old-implement "workshops," that in one of the tools used at Brandon we have the lineal descendant of one used by the old flint-workers. The flakes from which the arrow-heads and some other implements were made had a single back, that is, one rib running down the centre of the back. This can be still traced on many of the worked implements, and the waste and many of the unused flakes which still lie about in thousands are of this character. (See Fig. 17.) Now these flakes were struck with a round or ovoid pebble, of which many specimens, showing the battered ends, have been found. The old English gun-flints were also made from single-backed flakes, or from flakes which had no ribs whatever, and these were struck with an oval hammer. The backless English flakes, and many of the single-backed ones, are still struck with the ovoid English hammer. It is well known that the shape of stone tools, such as celts, were copied in bronze, and bronze patterns were used in casting iron tools. This standing over of old habits is a very common and striking phenomenon, and obtains not merely in the shape of tools, as may be seen in the gauge of railway lines, which is the breadth of the old coaches, and in a still more singular case in the aversion from the use of horse-flesh by the English, this article of diet having been prohibited in early Christian times because it was eaten in the feasts to Odin. Such being the case, it seems to me to be highly probable that the old English flaking-hammer was a metal copy of the old stone-fluker. True, the modern tool is perforated and hafted, while the implements known as hammer-stones were merely grasped in the hand, and doubtless this was the first method pursued; but ovoid round-faced perforated celts are by no means unknown, several having

---

* It is said in Brandon that at the commencement of this century the gun-flints were better made than at present; but this is at variance with the opinion I have formed from an examination of quantities of old flints still in store at the Tower.

† The moderns, however, excel in flaking.

been found in this neighbourhood. These, which Mr. Evans describes together in Chap. ix. of his classic work on stone implements, have had various uses ascribed to them. "By some
" antiquaries these perforated pebbles have been regarded as
" weights for sinking nets, or for some such purpose; but in
" most cases this is, I think, an erroneous view—firstly, because
" the majority of these implements show traces, at their extre-
" mities, of having been used as hammers; and, secondly, because,
" for such purposes as weights, there can be no doubt that
" the softer kinds of stone, easily susceptible of being pierced,
" would be selected, whereas these perforated pebbles are
" almost invariably of quartzite or some equally hard and rough
" material.

" There are some instances, indeed, in which the perforation
" would appear to be almost too small for a shaft of sufficient
" strength to wield the hammer, if such it were; but even in such
" cases, where hard, silicious pebbles have been used, they must
" in all probability have been intended for other purposes than
" weights."*

Now any one of these would make an admirable flaking-hammer, and the small "eyes" so far from offering the same difficulty on this explanation as on the supposition that they were ordinary hammers, is evidence in favour of it, because the blow given in flaking is not a heavy one, indeed in most cases the hammer is merely allowed to fall upon the stone, and the perforation or " eye" is always made as small as possible. The principal desideratum in a flaking-hammer is to get the greatest possible weight with the least possible size; hence the eyes are made so small that they are only sufficient to carry the handle : if they could be hafted as conveniently they would not be perforated at all. Here then I would look for the immediate predecessor of the English flaking-hammer, and if the deduction be accepted it is another point in favour of the supposition that the working of flint has gone on continuously since neolithic times.

Slowly as the flint-trade has progressed there are a few changes we can point to as having occurred in modern times. Of these the oldest and most important was the introduction of the present flaking-hammer, which at once gave rise to the more useful and elegant double-backed flints. Another change was the abandonment of the collar to the knapping-hammer by the present generation, a course which shows enhanced skill in cutting the flint so that it does not fly towards the knapper. Within the last five or six years the practises of damping the stone to lay the dust has become more general, showing a greater consideration for health, of sloping the stake to make the position of the left hand easier, and the substitution of new for old leather for the stage, have greatly increased. Such changes show how gradually an industry is modified, or, as it is becoming the fashion to express it, " evolved."

---

* Evans, Anc. Stone Imps., p. 194.

The evolution of the gun-flint from the strike-a-light trade has been already discussed, and no one can compare the strike-a-lights in the collection with the French gun-flint (see Figs. 59 and 60), or the French strike-a-light (Fig. 21) with the same specimen without being struck with the probability of the suggestion. But these strike-a-lights and old gun-flints are singularly like some of the ancient "scrapers," as pointed out by the astute author of Stone Implements, who further remarks, "I find, more-
" over, that by working such a flint and a steel or *briquet* together,
" much the same bruising of the edge is produced as that apparent
" on some of the old "scrapers." I come, therefore, to the con-
" clusion, that a certain proportion of these instruments were in
" use, not for scraping hides like the others, but for scraping iron
" pyrites, and not improbably, in later days, even iron or steel for
" procuring fire."* This is sound reasoning, and if the form of strike-a-lights *has not varied since then*, my suggestion of the origin of the English hammer is not very far fetched.

I am fully convinced that in the Brandon gun-flint trade we have a relic of the true stone age, and when philologists attack the question it is probable that some of the curious words in use among knappers, which I have carefully preserved in this work, may be found to be outliers of the old pre-Aryan tongue.

The value of the study of gun-flints to geologists is greater than has been supposed, for it affords the best experience of the fracture of flint, and at once enables us to determine the natural or artificial nature of the fracture. But it does more, it teaches us to distin-guish many individual features of the flint, and of the tool with which it is struck.

Those whose studies have not led them to pay particular atten-tion to flint often express doubts as to the artificial nature of some of the ruder implements, and, still more frequently, of flakes. Geologists even have been heard to declare that some of these uncouth specimens could be matched from any gravel heap. To such I would reply that if they will conscientiously examine care-fully a gravel pit and handle the stones, they will find that such is not the case, and the longer they search the stronger this opinion must grow, until, eventually, if any doubt remain it will rather be as to whether the specimens in question were ever found in gravel, so rare are they. If by good fortune he find one he will then assuredly treasure it as a peculiarity, even if he still doubt its human workmanship. I am perfectly convinced that it is as easy for an experienced observer to discriminate between artificial and natural flint chips, as it is for the ordinary observer to dis-tinguish stone arrow-heads from rifle-bullets. This is in conse-quence of the very peculiar manner in which flint and some other rocks fracture. *It requires a sharp rebounding blow delivered with a definite amount of force to detach a flake.* If the blow be dead the stone is only bruised; if it be bounding, but too slight, the bruising shows an incipient cone, but no flake is struck; if, again,

* " Ancient Stone Implements," p. 283.

the blow be too heavy the stone is shattered. Moreover, the blow
must be delivered in a certain direction. Now it is possible
that all these conditions will be found sometimes in nature, as
when a stone is flung upwards by a torrent or wave and falls upon
a larger stone or rock-surface. But it is clearly *impossible* that
they shall occur in the majority, or even in any considerable per-
centage, of cases. Hence, if we find a single flake showing the bulb
of percussion, we cannot say for certain that it is artificial (unless
it is one of a series of flakes, in which event the shape is modified) ;
but if we find a number of them in a limited area we may be quite
sure that they are the work of intelligent beings.

If, again, we find a stone which shows that several flakes have
been removed we may be certain that they have been struck
intentionally, for the chances are infinitely against a single stone
having been several times placed by nature under the requisite
conditions. Besides, these flakes have all been removed at the
same time, as shown by the condition of the surface, and this
again makes the chances less in favour of nature. Finally, it can
invariably be shown that the flakes have been removed with a purpose, and it is always easy to show (1) whether the flakes were
struck to be worked up into arrow-heads, &c.; or (2) whether they
have been removed so as to bring the core to a definite shape ; or
(3) whether the core, after useful flakes have been removed, has
been worked up again to subserve some useful purpose. Again,
it is always possible to tell whether a tool has been re-chipped
along the edge after having been worn down by use ; and if a tool
be broken, whether the fracture is of the same age as the tool, or
subsequent to its manufacture. Lastly, when unfinished tools are
found, as is frequent on the sites of the old workshops, the reason
for their rejection can always be determined, and I have found
more than one flint celt which, after being begun, has been rejected
for some flaw but used as a hammer, and the blows administered
as a hammer are always perfectly distinct from those given in
shaping the implement.

I have carefully examined upon the ground a very large
number of ancient flakes and cores, and have had the advantage of
the experience of Mr. Southwell, than whom none is better capable
of judging of these things, and I find that the outside, waste, and
useful flakes can be distinguished, and, moreover, different peculiarities of fracture which mark the skilled from the unskilled flaker.
Nearly all these seem to have been struck with a rounded tool, and
the flaking hammers must have varied in size from very large ones
to very small. Most of the flakes are without backs, such as those
generally struck with an English hammer; many are single-
backed, and a few double or triple-backed. These latter, however,
are certainly accidental, having arisen in two ways; namely, either
from a large flake covering the faces of several flakes previously
removed, having been removed to obtain a fresh smooth surface, or
from a small flake having been taken off a large single-backed one,
thus leaving two ribs on a large thin surface. This latter seems
to have been a usual practice, and I have specimens showing

abortive attempts. to remove this small flake. A series of specimens illustrating the different stages of the old manufacture is in
process of formation, and will be deposited in the museum as soon
as I have completed it.

Anyone who will carefully study the present collection, will be
able to distinguish the points I have described, and that the individual peculiarities of the flints may be readily observed, I have
given short notes of each gun-flint, and also showed wherein it
differs from what may be called the ideal types. Although these
are described as flaws, they must not be considered as evidence of
inferior workmanship, for the faults lay in the stone, and not in
the knapper, who is one of the most skilful in Brandon.

DESCRIPTION OF SPECIMENS.

Flint.

1. Quartered floor-stone. The entire mass shows the average
size of the flints as delivered to the knappers. It is broken into
five quarters, and the positions of the blows are numbered in their
order of succession; the arrows point to the places of percussion.
No. 4 has broken badly, and the coat has been accidentally
chipped. This specimen well illustrates the difference in thickness
between the top and bottom coats.

2. A specimen of floor-stone showing the double-coat.

3. A "pot-lid," or natural fracture. One side of the inner
surface shows what is termed a "frosty face."

4. Flint from Icklingham, showing bulb of percussion. Very
fine stone.

5. Outside waste flakes, called *shives*, showing the coat. From
old deep pits at Icklingham. Very good stone, superior to any
now dug.

6. Single and double-backed flakes. Figs 16 and 17.

7. Old English flake.

8. Flake marked to show how it will knap; the pieces will form
waste, one horse-pistol, three carbines, and one double. Fig. 19.

9. Flake cut into five pieces to make four gun-flints.

10. Four gun-flints made from one flake.

Tools.

11. Stone-digger's pick. Has been in use. Fresh pointed for
work.

12. First quartering-hammer, new. See p. 16.

13. Ditto, old, see p. 16.

14. Second quartering-hammer, new. See p. 16.

15. Ditto, old. See p. 16.

16. English hammer. See p. 18.

17. First flaking-hammer, new. See p. 17.

18. Ditto, old. See p. 17.

19. Second flaking-hammer, new.   See p. 17.
20. Ditto, old.   See p. 17.
21. Flaking candle-stick.   See p. 21.
22. Knapping hammer, new.   See p. 19.
23. Ditto, old.   See p. 19.
24. Knapping candle-stick.   See p. 21.
25. Model block, with stake and stage, full-sized ; block about one sixth natural size.   See p. 19.
26. Stake, full-size.   See p. 20.
27. Stake leather.   See p. 20.
28. Staging wood.   See p. 20.
29. Knapping leather, upper surface marked.   See p. 20.
30. Wringings.

Gun-Flints.

Numbered successively 1 to 33.   Two points require special mention ; the first is that no gun-flint is made by measure, but all by the eye; the second is that the remarks attached to each specimen only show wherein they differ from an ideal type, and do not denote inferior workmanship, for they are exceptionally well made.

It will be noticed that the distinction between *bests* and *seconds* is arbitrary, and that good seconds may be as well made in every respect as inferior bests.   As a rule, however, seconds are thinner than bests, or as thick but of inferior workmanship or material.   Commons are single-backed, but may be as well made as bests.   Double-edged flints are longer than single-edged of the same kind, because they have no heels to be cut off.   No. 1 among the second singles is in reality a super-single, but it is left in its present condition because it was placed there by the knapper, and serves to illustrate how bests may be picked out from a mass of seconds.   The dimensions given are in inches.

The terms applied to the parts of a gun-flint are illustrated by the following cuts and descriptions :—

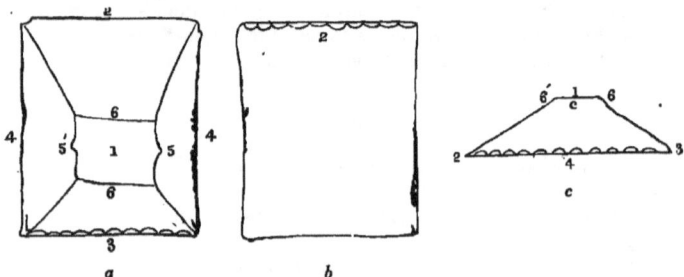

Fig. 25.—*Back, face, and side view of best Carbine.*

(*a*) Back view ; (*b*) face view ; (*c*) side view ; 1, back ; 2, edge ; 3, heel ; 4, 4, sides ; 5, bulbs or " knots ; " 6 ribs.   The sides, edge, and heel are slightly hollowed, or waisted, which is considered a point of beauty.

1. Wall Pieces.

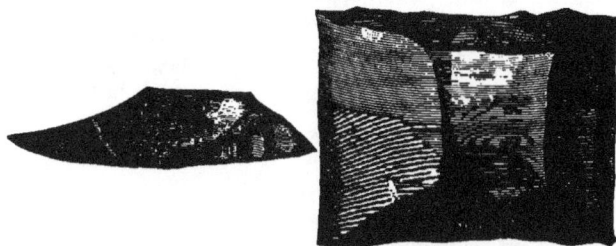

Fig. 26.—*Wall Piece.*

Size.—2·0 × 1·5 × 0·5.

Make.—Single-edged, rudely made, large flints; very variable in size.

2. Large Swan.

Fig. 27.—*Large Swan.*

Size.—1·5 × 1·2 × 0·3.

Make.—Back nearer heel than edge, moderately broad; good thick flint.

1. Spotted flint, back very broad. Side-bulbs opposite each other, but unequal in finish.

2. Good square back, nearer heel than front, side-bulbs opposite. A very well made flint.

3. Back not so square as No. 2, side-bulbs not opposite, edge long and good.

4. Back moderately good, side-bulbs not opposite, very long front. Too thin to be a good flint.

5. Solid flint not clean cut, bulbs not opposite.

6. A poorly made flint of inferior material.

7. Very good black flint, but spotted on the edge, side-bulbs not opposite, heel-bulb prominent.

8. A well-made flint with good back, side-bulbs not opposite.

9. Narrow back, and consequently long front and short heel. Heel spotted, and the unequal hardness thus caused has prevented its being clean cut. Side-bulbs not opposite.

10. Beautifully-made flint of good material, back central but rather too narrow. Bulbs opposite, sides equally well cut.

Note.—If these flints were assorted, Nos. 2, 3, 5, 7, 8, and 10 would be "bests;" Nos. 1, 4, 6, and 9 "seconds;" Nos. 1, 7, and 9 might go as "spotted" or "mixed" flints. No. 10 would have made a double-edge, but such are never made in this size.

3. Best Musket.

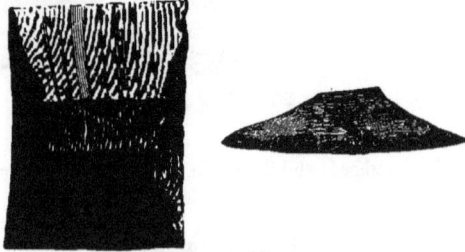

Fig. 28.—*Best Musket.*

Size.—1·3 × 1·1 × 0·4.

Make.—Back nearer heel than hedge, moderately broad ; good thick flint.

1. Beautifully made flint, with good square back, rather too central, side-bulbs opposite and in centre of side, sides well cut, heel and edge quite straight.

2. Almost as well made, but back not so square, side-bulbs opposite and in centre of back, trace of heel-bulb.

3. An excellent flint, with back squarer even than No. 1, but not quite such good flint ; side-bulbs opposite and in centre of side ; all the workmanship is perfect.

4. A well-made flint, with back pretty square, side-bulbs opposite and in centre of side.

5. A good flint, with back rather irregular, side-bulbs nearly opposite, right side not so cleanly cut as left.

6. A good flint, with irregular back, side-bulbs nearly opposite, sides equally clean cut, but heel not so.

7. Back irregular, side-bulbs not opposite, heel not clean cut.

8. Irregular back, side-bulbs not opposite, heel not clean cut, and its bulb showing.

9. A good flint, with the most irregular rib in front, side-bulbs not opposite. Thinner than it ought to be, and back too narrow.

10. A well-made flint, but too thin, and back too narrow ; side-bulbs not opposite.

Note.—Nos. 9 and 10 might equally well be called Second Muskets in virtue of their thinness. The back should not be central, but nearer the heel. It is noticeable here, and indeed generally, that one side of a flint is often better made than the other ; the better one being the first made. Flint, after being handled, never works so well as it would have done before.

4. Second Musket.

Fig. 29.—*Second Musket.*

Size.—1·3 × 1·1 × 0·3.

Make.—Second Muskets, when specially made, are thinner than Best Muskets, but a best with a flaw goes in as a second.

1. Was made for a best, but put in with the seconds because it was not square enough ; otherwise it is a very good flint, and superior to Nos. 9 and 10 of the bests.

2. A very good typical second, beautifully made in all respects except a slight notch on the right side; but for this and its thinness it would have been a best.

3. Thick and well-made enough for a best, but with a notch on the right side.

4. Well-made enough for a best, for which it was intended, but put into seconds because it turned grey, and is wrung or twisted. Whenever the edge is wider than heel the flint goes as a seconds.

5. Is an inferior quality, brackly and hollow-fronted.

6. Wrung flint ; heel longer than edge.

7. Well-made flint, but a second for similar reasons to No. 4.

8. One side badly made in consequence of the flake being thickest there.

9. Inferior edge ; not square.

10. Inferior edge wrung.

5. Double-edge Musket.

Fig. 30.—*Double-edge Musket.*

Size.—1·3 × 1·2 × 0·4.

Make.—Slightly longer than other muskets, but with two edges, and consequently no heel.   The back and bulbs are central and the flint of good consistence.

1. Well-made flint, but rather too thin, and the back rather too broad, and not quite central.

2. A better made flint, but not quite pure in colour; bulbs nearly opposite, and central.

3. Well-made ; back not quite central.

4. One edge gapped, back too broad, and the best edge too short.

5. Wrung; too thin ; one edge too short.

6. Wrung ; one edge gapped; back nearly perfect, but not quite central.

7. Too thick ; back too narrow ; edges not square.

8. Similar to No. 7.

9. Inferior back and side.

10. Good edges, but back too wide, and one edge too short.

6. Common Musket.

Fig. 31.—*Common Musket.*

(The apparent double-back is the effect of a flaw in the flint.)

Size.— 1·3 × 1·1 × 0· 4.

Make.—Like other muskets, but single-backed.   Never made double-edged ; hence the heel should be short.

1. An excellent flint ; square; long fronted ; good thickness ; bulbs opposite and on back.

2. Well made, but the flint has not broken quite cleanly; the right side shows well the difference in make of the two sides of a flint.

3. One side bad ; bulbs not opposite.

4. Not pure in colour : back too central.

5. „   „   too thin.

6. Thin ; back too central.

7. Good flint, but edge gapped and the " chalk " showing on the back.

8. Bad back.

9. Too thick ; back not straight.

10. Inferior material, shows the grisly fracture on the heel.

7. Mixed Grey or Spotted Musket.

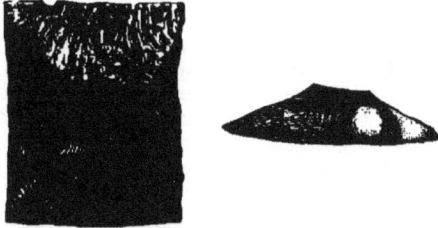

Fig. 32.—*Mixed Grey or Spotted Musket..*

Size.—1·3 × 1·1 × 0·4.

Make.—Same as other muskets, but of a spotted flint. None of these, from the nature of the stone, are so cleanly cut as black flints.

1. Excellent flint, but with edge gapped.
2. Not quite so good, ditto
3. Not so cleanly cut, but with better edge.
4. Well made, but wrung.
5. Good edge, but not square.
6. Well made, but back unsymmetrical, and too wide.
7. Back too central, edge gapped, too thin.
8. Not square, ditto ditto
9. Unevenly cut, bad edge.
10. Well made, but much wrung.

8. Solid Grey Musket.

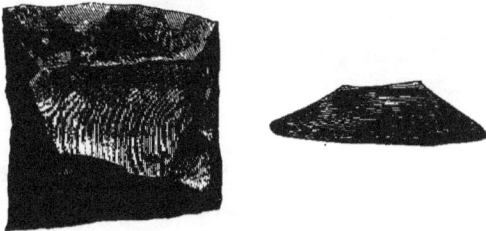

Fig. 33.—*Solid Grey Musket.*
(Icklingham make.)

Size.—1·3 × 1·1 × 0·4.

Make.—Same as other muskets, but of grey flint, which cuts grisly.

These flints vary in thickness greatly; the above size being the theoretical one, the length varies from 1·3 to 1·4, the width from 0·9 to 1·1, the thickness from 0·3 to 0·5.

1. Well made for a grey, back not good ; too thin.
2.   .Ditto, but back too broad.
3. Sides and heel rough, too thick.
4. Good back, fair sides, face wrung.
5. Well made, but single backed.
6. Good edge, but single backed and wrung.
7. Fairly good, but with uneven back.
8. Too thin, back much too broad.
9. Too narrow and thick, wrung.
10. Coarse flint, single-backed and irregular.

9. Chalk-heeled Musket.

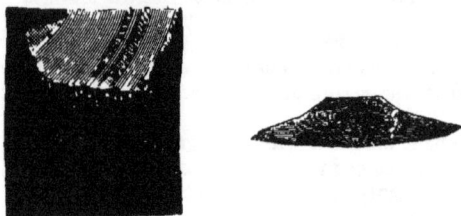

Fig. 34.—*Chalk-heeled Musket.*

Size and make like other muskets, but purposly showing the coat or "chalk" on the heel. The three specimens in the collection are beautifully made in all respects.

10. Best Carbine.

Fig. 35.—*Best Carbine,*
(The draughtsman has omitted to show the trimmed heel, so that it looks like a double-edged flint.)

Size.—$1.2 \times 1.0 \times 0.25$.
Make.—A single-edged flint, like a musket but smaller.
1. Excellently made; back almost perfect, but bulbs not quite opposite.
 2. Very good, but the back rather too broad, and the bulbs not quite opposite.
3. Good flint, but back not quite square.
4. Very good flint, slightly waisted, and heel rather too narrow; bulbs not opposite.

5. Good flint, with edge gapped; bulbs not quite opposite, and back too broad.

6. Good flint; sides not equal, and one with two bulks.

7. Well made waisted flint, but with edge gapped, and back not quite in right place.

8. Well made, but wrung.

9. Too thick.

10. Ditto, but well made; edge gapped.

11. Second Carbine.

Fig. 36.—*Second Carbine.*
(Heel not drawn.)

Size.—1·2 × 1·0 × 0·25.

Make.—Made for bests, but condemned for some flaw which is alone pointed out in the following list. Some might go as bests, as Nos. 9 and 10 of the bests might have been put in as seconds.

1. Flawed face; "chalk" shows on back.
2. Wrung.
3. Imperfect edge and heel.
4. Not square, and one side too straight.
5. Bad colour.
For make any of the above might have gone in as bests.
6. Inferior back and edge.
7. Too thin.
8.   „   bad back.
9.   „   „
10. „   „

12. Carbine, Double-Edge.

Fig. 37.—*Double-edged Carbine.*

Size.—1·2 × 1·1 × 0·25.

Make.—Like other carbines, but with the back central so as to make the two edges of equal length.

1. Excellently made, the only fault being a thickening of one end of an edge, but this is of no moment.

2. Equally well made, but one edge gapped.

4.   „   but back rather too broad.

5. Good, but slightly wrung.

6. One edge gapped; back too narrow.

7.   „   „

8.   „   back too narrow.

9.   „   „   not central.

10.   „   too thin; back too broad.

The whole of these are admirably made.

13. Common Carbine.

Fig. 38.—*Common Carbine.*
(The trimmed heel not shown.)

Size.—1·2 × 1·0 × 0·25.

Make.—Like other single-edged carbines, but single-backed.

1. Excellent, but with a gapped side.

2.   „   but back too central.

3.   „   edge not quite so good as above.

4. Not quite so well made.

5. Very good, but gapped edge.

6.   „   but too thin.

7. Back too central, spotted on face.

8. Too thin.

10. Inferior flint.

14. Common Carbine. Double-Edge.

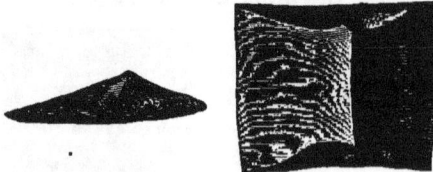

Fig. 39.—*Common Carbine. Double-edged.*

Size.—1·2 × 1·1 × 0·25.

Make.—Like No. 12, but double edged; hence the back is central.

1. Well made, but rather thin.
2.   ,,   edge not true.
3.   ,,   thin.
4.   ,,   one side grisly.
5.   ,,   back not straight.
6.   ,,   not cleanly cut.
7.   ,,   bulbs not on rib.
8.   ,,   one edge not true.
9. Back wrung, edge gapped.
10.   ,,   ,,

15. Grey or Spotted Carbine.

Fig. 40.—*Grey Carbine.*

Size.—1·2 × 1·0 × 0·25.

Make.—Like other single-edged carbines, but of grey or spotted flint, and hence ruder in workmanship.

1. Very good flint for spotted stone.
2. Equally well made, but back too long.
3. Edge gapped.
4. Very good, but rather narrow.
5. Good, but wrung.
6. Wrung and small.
7. Irregular backed.
8. Wrung and poor backed.
9. Well made, but thin.
10. Wrung, coarse stone.

16. Chalk-heeled Carbine.

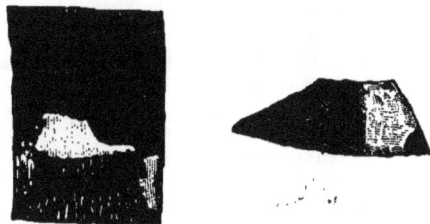

Fig. 41.—*Chalk-heeled Carbine.*

Size.—1·2 × 1·0 × 0·25.

Make.—Like other single-edged carbines, but with the coat or " chalk" purposely showing on the heel.

Two specimens are shown, marked 3 and 4 ; the former is well made, and the latter would be equally good, but that the flint shows a tendency to double-coating.

17. Best Horse Pistol.

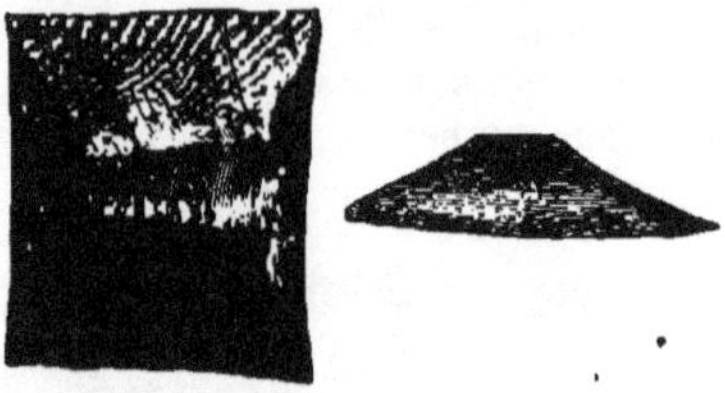

Fig. 42.—*Best Horse Pistol.*
(Heel not shown.)

Size.—1·1 × 0·9 × 0·3.
Make.—Squarer than carbines, and smaller.
1–5. Perfect as flints can be.
6. Very good, edge slightly gapped.
7. Back too central.
8.   ,,   narrow, edge gapped.
9. One side too straight.
10. Inferior material, edge too short; it had to be re-chipped.

18. Best Horse Pistol. Double-Edge.

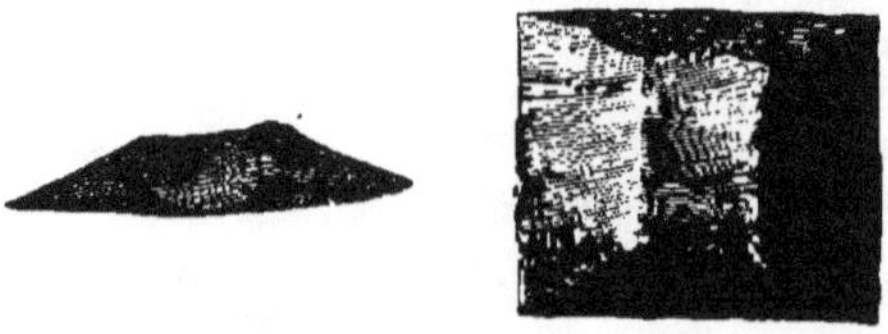

Fig. 43.—*Best Horse Pistol.   Double-edged.*

Size.—1·1 × 1·0 × 0·3.
Make.—Similar to other horse pistols, but with back central to make the edges equal in length.
1. Excellent flint, but edge gapped, and one side too straight.
2.   ,,   but one edge not quite perfect.
3.   ,,   ,,   ,, gapped.
4.   ,,   but thin.
5.   ,,   ,, rather short.
6.   ,,   ,, thin.
7. Back too broad, edge gapped.
8.   ,,   ,,   too thin.
9.   ,,   ,,   ,,
10. Thin.

19. Second Horse Pistol.

Fig. 44.—*Second Horse Pistol.*
(Heel not shown.)

Size.—1·1 × 0·9 × 0·3.

Make.—Same size as bests, but rejected for some flaw such as thinness; these flaws are enumerated below.

1. Rather narrower in the heel than edge.
2.   „   „   edge „ heel.
3. Inferior heel.
4. Wrung.
5. Gapped edge.
6. Wrung.
7. Thin.
8.   „
9.   „
10. Back too thin.

20. Second Horse Pistol. Double-Edge.

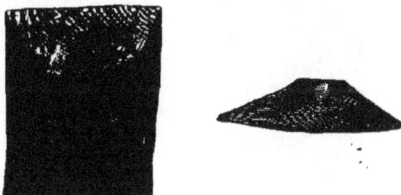

Fig. 45.—*Second Horse Pistol. Double-edged.*

Size.—1·1 × 1·0 × 0·3.

Make.—Like other Second Horse Pistols, but double-edged, which brings the back to the centre.

1. Nearly perfect flint; would be a best if thicker.
2. Rather short; back narrow.
3. Short, back not central.
4. Very nicely made but thin.
5. Good flint but short.
6. Not quite square.
7. Wrung.
8. Too thin.
9. Not good colour; thin.
10. Edge gapped.

21. Common Horse Pistol, or Large Common Gun.

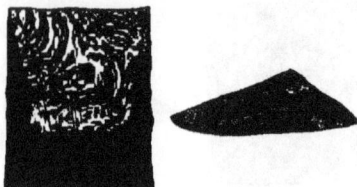

Fig. 46.—*Common Horse Pistol.*

Size.— 1·1 × 0·9 × 0·3.
Make.—Like other horse pistols, but single backed.
1. Very good, but bulbs not opposite.
2.   „   „
3.   „   back not level.
4.   „   bulbs opposite; rather irregular edge.
5.   „   thin.
6. Well made, but thick.
7. Wrung.
8. Gapped edge, wrung, spotted back.
9. Wrung.
10. Poor back.

22. Common Horse Pistol. Double-Edge.

Fig. 47.—*Common Horse Pistol. Double-edged.*

Size.— 1·1 × 1·0 × 0·3.
Make.—Like common horse pistol, but double-edged, hence the
back is central.   These are only made for special orders.
1. Well made, but one bulb not on back.
2. Bulbs right but edges not so good as No. 1.
3. Bulbs not opposite, one edge slanting.
4. Too thick.
5. Very nicely made, but rather small.
6. Too thin.
7. Thin and short.
8. Thin, but very nicely made.
9. One bad edge.
10. Wrung.

23. Mixed Grey or Spotted Horse Pistol.

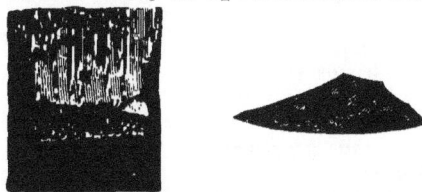

Fig. 48.—*Mixed Grey Horse Pistol.*

Size.—1·0 × 0·9 × 0·3.
Make.—Like others, but of spotted flint.
1. Nicely made, bulbs not quite opposite.
2.   „   edge gapped.
3.   „   rather thin.
4.   „   „
5.   „   back too central.
6. One side inferior.
7. Wrung.
8. Longer than the others.
9. Back too broad.
10.   „   narrow.

24. Chalk-heeled Horse Pistol.

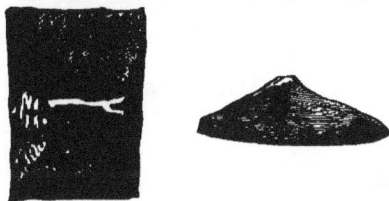

Fig. 49.—*Chalk-heeled Horse Pistol.*

Size.—1·0 × 0·9 × 0·3.
Make.—Like others, but purposely showing the coat or " chalk "
on the heel.
5. Well made.
6. Short.
7. Short-heel.
8. Single-back.
9. Poor back.

25. Super Single.

Fig. 50.—*Super Single.*

Size.—1.0 × 0·85 × 0·2.
Make.—Similar to, but smaller than horse pistols.
1. Excellently made.
2. „ bulbs not quite opposite.
3. „ edge not quite true.
4. „ back not so good as others.
5. „ too thin, gapped edge.
6. Thin but good.
7. Very nicely made, bulbs pert, back too broad.
8. Back narrow.
9. „ broad.
10. „ „

26. Second Single.

Fig. 51.—*Second Single.*

Size.— 1·0 × 0·85 × 0·2.
Make.—Same as super singles but rejected for some flaw as detailed below.
1. A singularly perfect flint in every respect. Has got in here by mistake. It is a super single.
2. Gapped edge.
3. One side too straight.
    Any of the above would go as super singles.
4. Thin, broad back.
5. Poor edge, broad back.
6. Thick „
7. Thin, broad back.
8. Small „
9. Poor edge and back.
10. „ broad back, thin.

27. Fine Single, or Small Common Gun.

Fig. 52.—*Fine Single.*

Size.—1·0 × 0·8 × 0·2.

Make.—A trifle narrower than super single.
1. Well made, but rather thick.
2.   „   bulbs not opposite.
3. One side too straight   „
4. Unsymmetrical.
5. Gapped edge.
6.   „
7. Back too broad.
8.   „   thin.
9.   „   „
10.   „   „

28. Super Double.

Fig. 53.—*Super Double.*

Size.—1·0 × 0·7 × 0·25.
Make.—Between horse pistol and single in thickness, same length as, but narrower than single. Made from a wide flake, with the heel and edge cut cleanly off.
1. Beautiful edge, but sides not equal.
2. Very nicely made, bulbs perfect.
3. In make quite perfect, but wide enough for a single.
4. Very good, heel short.
5. Edge gapped.
6. Back narrow.
7. Thin.
8-10.   „
    The whole of these are remarkably well made.

29. Second Double or Rifle.

Fig. 54.—*Second Double.*

Size.—0·95 × 0·65 × 0·2.
Make.—Same as super double, but condemned for some flaw, as recorded below.
1. Edge slightly rough.
2. Rather thin.
3. One side too straight.
4. Narrow back.
    All the above would do for super doubles.

38856.   E

5. Too square.
6. „ thick.
7. „ thin, inferior side.
8. „ „ back too broad.
9. Small.
10. Gapped edge.

30. Fine Double.

Fig. 55.--*Fine Double.*

Size,—1·0 × 0·65 × 0·23.
Make.—Like super double but narrower. Practically most of them are scarcely distinguishable from the super doubles.
1. Excellent flint.
2.   „
3.   „   edge gapped.
4. Well made, but sides unequal.
5.   „   but wrung.
6.   „   large.
7.   „   „
8.   „   back too broad.
9. Edge too short.
10. Too short.

31. Super Pocket Pistol.

Fig. 56.—*Super Pocket Pistol.*

Size.—0·75 × 0·65 × 0·2.
Make.—A small square flint, the same size as a rifle, but shorter
1. Excellently made in all respects.
2.   „
3.   „
4.   „
5.   „ but rather broad.
6. Heel short; flint too long.
7. Poor colour.
8. Back and edge not perfect.
9. Thin.
10. Back too broad.

32. Fine Pocket Pistol.

Fig. 57.—*Fine Pocket Pistol.*

Size.—0·75 × 0·6 × 0·2.

Make.— Same as supers, but narrower, and often of inferior make, whence they might be called seconds.

1. Excellent flint.
2.   ,,
3. Thin.
4.   ,, but broad back.
5. Bad back, too narrow.
6. Too short.
7. Too thin, back too broad.
8. Short.
9. Poor.
10. Thin, broad back.

33. Old English Gun-Flints.

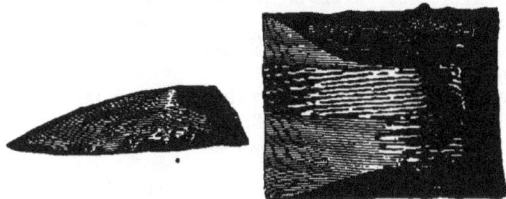

Fig. 58.—*Old English Gun-Flint.*

Made from English flakes.

Fig. 59.—*French Gun-Flint.*

French gun-flints of two sizes. Over 100 years old. . Their identity of shape with strike-a-lights, and some old " scrapers " is very patent. See pp. 43 and 44.

Fig. 60.—*Cut and Polished Flint.*

E 2.

Cut and polished flint. Made for experiment 30 years ago, but never in demand. The cost price was 9*d*. each.

Polished French gun-flints made at the same time.

Flint Locks, &c.

Double-barrelled old flint-lock pistol of very fine workmanship, with flint inserted. This weapon takes a pocket-pistol size.

Old carbine flint-hammer with flint inserted.

Flint fired in the above pistol 100 times to show the manner of wearing. See p. 4.

Ditto, showing how a flint wears if too wide for a weapon. See p. 4.

Waste.

Flaker's chips.

Knapper's chips. These specimens show how readily distinguishable are artificial from natural chips.

Miscellaneous.

Strike-a-lights. These are shown *A*, of ordinary shape, like large French, or old English gun-flints; *B*, horse-shoe shaped; *C*, straight-sided round-edged; *D*, half-round; *E*, circular. All of these are at present made for sale, and can be matched among neolithic so-called "scrapers." They are made generally from English flakes; but (when ordered) any flake, English, single, or double-backed, is used, provided it is thick and broad. See p. 36.

Discs, use unknown, made at Brandon.

Ditto, from Icklingham.

Square black faced-builder, with Knapper's initials, and date knapped upon the face. See p. 35.

Square, mixed-colour, faced-builder. •

Round, black, faced-builder.

Round, mixed-coloured, faced-builder.

Random, faced-builder, with " chip-back."

Rough builder.

Square, black faced-builder with initials, date, &c., like the Indian "snake pattern" agates.

Fig. 61.—*German Gun-Flint.*

ON THE AGE OF PALÆOLITHIC MAN,

I EMBRACE this opportunity of epitomising the evidence recently obtained by me, that there are three distinct horizons of beds yielding palæolithic implements. Of these, which I propose to call the *Early*, *Intermediate*, and *Late Palæolithic*, the last two are newer, and the first is older than the Chalky Boulder-clay. Combining the result of my own work in Norfolk and Suffolk with that of others, and taking into consideration also the researches of Mr. S. V. Wood, jun., in Yorkshire and Lincolnshire, the succession of deposits in the east of England formed during the great cycle of the glacial period seems to be as follows, in descending order :

1. *Plateau Gravels, &c.*—(Lincolnshire and Yorkshire.) Melting ice ; floods.
2. *Hessle Boulder-clay.*—(Lincsh. and Yorksh.) Not traced further south than the Lincolnshire border of the fens. Last ice-sheet.
3. *Gravel and Sand.*—Called in part Hessle Gravel by Mr. Wood. Mild period. Pleistocene fauna. Late palæolithic. Gravels of modern rivers, in part.
4. *Purple Boulder-clay.*—Not traced further south than North Lincolnshire. Large ice-sheet.
5. *Sands and Gravels.*—Mild period. Pleistocene fauna. Intermediate palæolithic.
6. *Flood Gravels* —In force in Norfolk and Suffolk. Formed by floods on melting of ice-sheet of Chalky Boulder-clay, and subsequently similar beds formed after deposition of beds 2 and 4. These are difficult to separate, and in this table are partly included in beds 1 and 3.
7. *Chalky Boulder-clay.*—Great ice-sheet. The most extensive of the glacial beds.
8. *Brandon Beds.*—Brick-earths hitherto only recognised in Norfolk and Suffolk. Pleistocene fauna. Early palæolithic. Boulder-clay sometimes underlies these beds. This may be an older clay, perhaps No. 11, but at present I incline to the belief that it most frequently consists of tongues of No. 7 intruded into and beneath No. 8.
9. *Sand and Gravels.*— Middle Glacial of Messrs. S. V. Wood, jun., and F. W. Harmer.
10. *Contorted Drift.*—Probably in part contemporaneous with No. 9. Generally consists of clays and loams much contorted.
11. *Lower Boulder-clay.*—Also known as Cromer Till. Apparently confined to the neighbourhood of the East Anglian Coast.

In passing from Yorkshire to Suffolk we thus travel in succession over four distinct boulder-clays, separated from each
other by beds of sand, gravel, and clay, which contain a pleistocene fauna.

It is a highly significant fact that beds containing this fauna
never lie upon the surface in the district occupied by the newer
boulder-clays Nos. 2 and 4. South of the R. Steeping these boulder-
clays are wanting, and there accordingly the beds containing the
pleistocene fauna begin to come on. This peculiarity of dis-
tribution receives a ready explanation, if we suppose the newer
ice-sheets to have more or less completely ploughed out the beds
in question from the northern area.

In the neighbourhood of Brandon, and indeed over the whole
of the midland and eastern counties, there are no ice-formed
deposits newer than the Chalky Boulder-clay.' In other words this
district has not been glaciated since the Chalky Boulder-clay was
formed. To this immunity from later glacial erosion we may
fairly ascribe the richness of its post-boulder-clay beds. But it
would clearly be erroneous to suppose that all the beds which, in
this area, overlie the Chalky Boulder-clay are post-glacial, since
they may belong to any part of the period shown in the table by
beds 1 to 8 inclusive.

In determining the age of deposits which in this district overlie
the Chalky Boulder-clay, we have two guides, namely, the palæontological evidence, and the physical. Now the palæontological
evidence is very strong. There are two distinct faunas, the one containing the remains of living species only, often associated with the
remains of neolithic man ; the other, rich in species either extinct
or no longer living in our latitudes, and frequently associated
with the relics of palæolithic man. When we critically examine
this latter fauna we find that while beds containing it are known
to pass under the Hessle and Purple Boulder-clays, no single
instance can be cited of their lying upon those beds, or even of
their occurrence at the surface in districts that have been over-
ridden by the ice which deposited those boulder-clays. It seems
therefore to be a legitimate inference that these beds are all of
older date than the Hessle Boulder-clay ; that they are, in fact,
inter-glacial, and not pre-glacial, or post-glacial. A physical
cause is thus given for the great break between the modern and
the pleistocene faunæ. This conclusion is further strengthened
by the fact that the mammalian fauna (which necessarily shows
greater change than the molluscan) is more closely allied to the
pre-glacial fauna of the old Forest-bed of the Norfolk coast, than
to that of the present time, or even of the neolithic period.

The physical evidence points the same way. We cannot apply the
test of superposition near Brandon, because the newer two boulder
clays are absent, but even if we confine our attention to the
palæolithic implement-bearing gravels of the present river-valleys
we have indisputable evidence of great changes having taken place
since the gravels were deposited, as was long since insisted upon
by Messrs. A. Tylor, Prestwich, and others. Further than this, I

have found that palæolithic implements (associated with the old pleistocene fauna) are not confined to the deposits in the present river-valleys, but are also found in gravels belonging to a prior drainage-system, which gravels cut across the present valleys. These are newer than the Chalky Boulder-clay. And still further, beneath the Chalky Boulder-clay itself, in a series of loams and sands to which I have given the name of Brandon Beds, palæolithic implements are found associated with the relics of extinct pleistocene mammalia.

Where palæolithic implements occur in deposits lying in very ancient valleys, it is clear that we cannot readily separate the remains of one period from those of another, for the containing material has accumulated steadily during a vast length of time, and moreover the material itself may have been worked over by the river more than once, and so have commingled relics of very different ages. But where we have to deal with a valley newer than the Chalky Boulder-clay the case is different: This is the case with that part of the Little Ouse valley from Thetford westwards; boulder-clay never lies in the valley, but the valley frequently cuts through boulder-clay. We are quite sure, then, that the implements found in the gravels of this part of the valley are newer than the Chalky Boulder-clay. To these I ascribe the name of *Late Palæolithic.*

There are, however, remains of a yet older valley system, which is also of newer age than the chalky boulder-clay. This old valley system has been traced by Messrs. Penning and Jukes-Browne in the district around Cambridge, and by myself in the vicinity of Brandon, where it yields flint implements. The deposits are gravels, which now cap the hills at a height of about 70 feet above the present river. They follow definite lines, and around Brandon are peculiar for the great quantity of quartz pebbles they contain. They run pretty nearly at right angles to the rivers Lark, Little Ouse, and Stoke, whose valleys indeed cut through them. They yield the well known "quartzite" implements as well as flint tools. Implements have been found at three places, Brandon Field or Gravel Hill, Lakenheath Hill, and Portway or Marroway Hill. To these I give the name of *Intermediate Palæolithic.*

These Intermediate Palæolithic beds are clearly much older than the Late Palæolithic, for they belong to a prior drainage system which ran at right angles to the present one, and at a considerable height above it. Nevertheless they are newer than the Chalky Boulder-clay, for they frequently repose thereon. We have given reasons for ascribing to the Late Palæolithic an age greater than that of the Hessle Boulder-clay, and it appears to me that in the absence of more definite knowledge, the facts are best explained by supposing the Intermediate Palæolithic beds to have been formed during the long interval which elapsed between the formation of the Chalky and Purple Boulder-clays. The early palæolithic, next to be described, are older than the Chalky Boulder clay, and if my suggestion be adopted, we have the following sequence of events:—

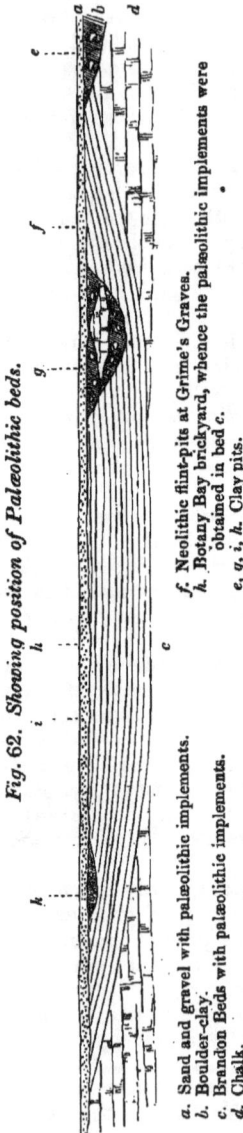

Fig. 62. Showing position of Palæolithic beds.

f. Neolithic flint-pits at Grime's Graves.
h. Botany Bay brickyard, whence the palæolithic implements were obtained in bed c.
e, g, i, h. Clay pits.

a. Sand and gravel with palæolithic implements.
b. Boulder-clay.
c. Brandon Beds with palæolithic implements.
d. Chalk.

1. Neolithic Period	-	- Post-Glacial.
2. Hessle Boulder Clay		- Glacial.
3. *Late Palæolithic*	-	Inter-Glacial.
4. Purple Boulder Clay	.	- Glacial.
5. *Intermediate Palæolithic*		Inter-Glacial.
6. Chalky Boulder Clay	-	Glacial.
7. *Early Palæolithic*		Inter-Glacial.
8 Lower Boulder Clay		- Glacial.

The only element of uncertainty in this table is the position of the Intermediate Palæolithic, No. 5. Of this we can as yet only say for certain that it is much older than the Late Palæolithic, No. 3, and newer than the Chalky Boulder Clay, No. 6.

The Early Palæolithic remains are found in a series of loams, sands, and gravels, overlaid by Chalky Boulder-clay. To this series I have given the name of Brandon Beds. They will be described at length in a separate work. They are very fragmentary, but seem to occur pretty nearly all over East Anglia wherever the Chalky Boulder-clay extends, always cropping out at or close to its base, *and never in a single instance occurring away from it.* This remarkable association is only explicable on the supposition that the Brandon Beds are older than the Chalky Boulder-clay, and indeed that clay can be actually seen lying thick upon them, and often contorting them, sometimes for a mile at a stretch. Up to the present time they have yielded implements or flakes at Botany Bay (near Brandon), Mildenhall Brickyard, High Lodge Mildenhall, Bury St. Edmunds, West Stow, and Culford. The first discovery was at Botany Bay, and at the time no boulder clay was visible at the precise spot, but it has since been met with, and I had the pleasure of the experience of Mr. Amund Helland when this fact was made clear. The following section, Fig. 62, shows the general lie of the beds (not to scale). It was drawn before I saw the boulder clay at Botany Bay.

At Mildenhall Brickyard and High Lodge good thick Chalky
Boulder-clay overlies the Brandon Beds whence many implements
have been obtained, and at Culford whence I dug out a good flake
in company with Mr. F. J. Bennett, the Brandon beds are worked
under 15 feet of Chalky Boulder-clay, and can be traced beneath
that deposit for the distance of a mile to the eastward.

In several places boulder-clay also underlies the Brandon beds.
This may in some cases be part of the Lower Boulder-clay, but I
believe in the majority of cases it is nothing but a tongue of
Chalky Boulder-clay intruded beneath the Brandon beds.

ON THE CONNECTION BETWEEN NEOLITHIC ART

AND

THE GUN-FLINT TRADE.

I think it can be confidently asserted that at Brandon we have,
as it were, an outlier of the Stone Age—that the flint-knappers
are the direct descendants of the old workers in stone, who dug
the ancient flint-pits at Grime's Graves, having preserved to this
day the method of mining, the shape of sundry tools, and the
peculiarities of certain flint implements.

Grime's Graves is the local name for an assemblage of rudely
circular depressions, varying in diameter from 30 to 60 feet, and
about 250 in number, which occur in a small fir-wood on the side
of a dry valley, in the parish of Weeting, about three miles north-
east from Brandon. One of these places was explored in the year
1870, and the conjecture that they were old flint workings
ratified. An admirable description of the exploration has been
communicated to the Ethnological Society by Canon Greenwell,
F.S.A., and appears in the second volume of their Journal. From
this paper many of the following notes respecting the Graves are
cited; where authorities are not given the observations are my
own. The Reverend explorer tells us that "the process [of
" working for flint] differs in some respect from that adopted by
" the present flint-raisers. The ancient workers sunk a circular
" shaft, gradually decreasing in size to the level of the stratum of
" the best flint, passing through the upper layer of the so-called
" wall-flint, but not removing any of that bed beyond what
" occurred within the limits of the shaft itself. When the floor-
" flint was reached, it was worked out to the extent of the pit;
" and then galleries were excavated in various directions upon the
" level of the bed of flint. In order that sufficient height might
" be obtained to enable the workmen to extract the flint, a con-
" siderable quantity of the overlying chalk has been removed, the

" galleries being on an average about 3 feet in height, though
" in some places the roof was 5 feet high. Their height,
" however, is very irregular, owing in some measure to the
" manner in which the chalk roof has given way in some places
" more than in others. In no case was any of the chalk below

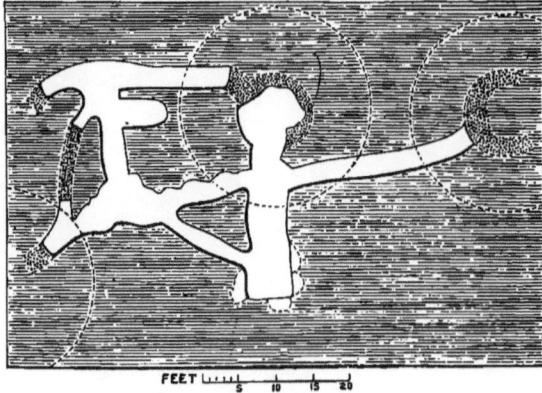

Fig. 63.—*Plan of Flint-Pit, Grime's Graves.*
(Reduced one half, from Canon Greenwell's Paper.)

The white portions show the extent of the excavations.
The dotted portions show the probable run of unexplored galleries.
The large circles show the approximate area of the top of the pits; and the small
ones of the bottoms thereof, as determined by Canon Greenwell. The pits are, in
reality, rudely rectangular.

Fig. 64.—*Corresponding portion to above of Modern Pit.*
(Drawn to the same scale.)

" the flint-bed removed,—a practice contrary to that of the present
" workmen, who, in making their galleries, excavate the chalk
" both above and below the flint. The galleries vary in width

" from 4 feet to 7 feet; and the flint was worked out
" beyond their sides as far as was practicable without causing the
" roof to give way. The position of the galleries will be better
" understood from the plan (Fig. 63), which shows their ramifica-.
" tions and the way they run into one another, than by description
" in words. . . . . . As one gallery was worked out, it
" was filled in again with the chalk excavated from other
" galleries, so that nearly the whole of them are now filled up with
" rubbish."*

In the above figures I have given, for the sake of comparison, a
reduction of Canon Greenwell's plan of the pit he opened at
Grime's Graves, and the corresponding portion of a modern flint-
pit. Before attempting to show the identity of the ancient and
modern processes, it is necessary to correct a slight error in the
above description. It is quite true that the present stone-diggers
burrow beneath the floor-stone. They still, however, work *above*
the flint in getting wall-stone, as in neolithic times, but never
"above and below the flint," as seems to be inferred by Canon
Greenwell.

The identity between the ancient and modern industries is
shown in several points. Wandering through the picturesque,
fern-clad wood in which Grime's Graves are situated, and then
passing out into the open heath of Broomhill, one comes upon the
site of modern flint-pits, no longer worked, which rival in number
their antique prototypes. It is impossible not to be struck with
the similarity of aspect between the two. The depressed basins
which mark the sites of their shafts; their close proximity to each
other, the heaps of chalk which surround them, can all be paralleled
in the wood hard by.

One striking difference has, however, been noted. The neolithic
pits, as shown in the plan, are said to be rudely circular, whereas
the modern shafts are rectangular. This description I find to be
deceptive. The "choldering-in" (as the diggers express it) of
the sides of the modern shafts has masked their angularity, and
they now appear almost as round as the neolithic pits. Moreover,
the only pit at Grime's Graves that was opened *positively shows to
this day traces of its original angularity* along the only exposed
chalk-face. If the remaining detrital matter were cleared away
this contour would be perfectly apparent. Even under present
circumstances I cannot understand how the true shape was
unnoticed. This "difference," then, does not exist.

In the extent of the burrows we find another strange connection.
I have already shown that their length is not arbitrary, but is
determined by the quantity of flint that can be got and carried
out in one day. The only burrow fully excavated at Grime's
Graves was 27 feet long, which is not far from the mean length
of a modern one. It is certain, however, that the stone-folk
worked in company in the same gallery, and not singly as now, as

---

* Rev. W. Greenwell, M.A., F.S.A. On the Opening of Grime's Graves in
Norfolk. Journ. Ethno. Soc., London, vol. ii., No. 4, 1871. Page 425.

is proved by finding several picks in a portion of a burrow of which the roof had fallen in while the workmen were away.* When, however, we recollect the inferiority of the ancient tools, and the extra labour attendant upon their method of digging, it is fair to suggest that one man now could perform as much work as two in those old times. It would seem, then, that the folks of yore, like those of to-day, carried out daily the products of their labour. However this may be, the similarity of extent of the ancient and modern burrows is a singular coincidence, that gathers weight in the presence of other facts of like nature.†

In modern pits stages are left at intervals of 5 feet, by means of which the digger gets to and leaves his work, and up which the stone and large chalks are carried. In Grime's Graves no such stagings were found, and Canon Greenwell is of opinion that they do not exist. Until the entire debris has been removed this cannot be asserted as a fact, though it is probably true. If so, we have here a decided difference between the two cases, but it is of little importance, for it would be difficult to say whether the hauling of the material by rude mechanical contrivances, was or was not more scientific than the manual labour of carrying the flint on the head up the artificial stages.

When we compare the ancient and modern mining plans, coincidences again crowd upon us. In working out floor-stone the moderns drive burrows, and do not merely undercut the flint as in digging for wall-stone: the ancients did the same. The moderns drive main- and side-burrows : so did the ancients. The moderns clear away the flint from semicircular openings in the burrows : so did the ancients. The essential features of a modern mine can be seen in the plan of Grime's Graves given above, which looks not very unlike a shockingly bad drawing of the modern plan. The latter, in fact, is an improvement upon the former, but in no single point does it bear the impress of originality.

The old pits were larger than the new ones at the top for the simple reason that more light was thus obtained, and as the means of illumination improved it is most likely that the diameter of the shaft diminished. There is, however, a great advantage in

---

* The men probably worked in pairs. Describing the above occurrence, Canon Greenwell says : "The roof had given way about the middle of the gallery, and " blocked up the whole width of it to the roof, On removing this, and when the " end came in view, it was seen that the flint had been worked out in three places " at the end, forming three hollows extending beyond the chalk face of the end of " the gallery. In front of these two hollows were laid two picks, the handle of each " towards the mouth of the gallery, the tines pointing towards each other, showing, in " all probability, that they had been used respectively by a right- and a left-handed " man. The day's work over, the men had laid down each his tool, ready for the " next day's work ; meanwhile the roof had fallen in, and the picks had never " been recovered. . . It was a most impressive sight, and one never to be for- " gotten, to look, after a lapse, it may be, of 3,000 years, upon a piece of work " unfinished, with the tools of the workmen still lying where they had been placed " so many centuries before." These picks still retained, upon their chalky incrus- tation, the impressions of the workmen's fingers ! Op. cit., p. 427.

† Canon Greenwell thinks all the pits communicated with each other. This, I cannot but think, is highly improbable, founded, as it is, upon a single observation. It would imply that a great many pits were worked simultaneously. Modern pits, as I have shown, sometimes communicate, but only when the stone is good.

burrowing under the stone. In working from above the burrow
must be at least high enough to enable the workman to stoop and
the labour of prising the stone from the floor is very great, whereas
in working from below the burrows need only be made high
enough for a man to recline upon his elbow, and thus much labour
in digging barren chalk is spared ; moreover the stone is easier to
get, inasmuch as when cracked its weight tends to bring it down.
Still this is probably only an improvement on the old system, for
it is natural that at first the diggers would not work deeper than
the stone they wished to raise.

A still more remarkable " coincidence " is found in the stone-
diggers' pick, and Canon Greenwell was struck with the similarity between the ancient and modern tools. He says, " The
" principal instrument used, both in sinking the shaft and in
" working the galleries, was a pick, made from the antler of the
" red deer, numerous examples of which were found in the shaft
" at various depths, and in the galleries. The pick, almost iden-
" tical in form with that, of iron and wood, used by the present
" workmen, was made by breaking off the horn, at a distance
" usually of about 16 or 17 inches from the brow end, and then
" removing all the tines except the brow tine."* Subjoined are
representations of an ancient and modern pick drawn to the same

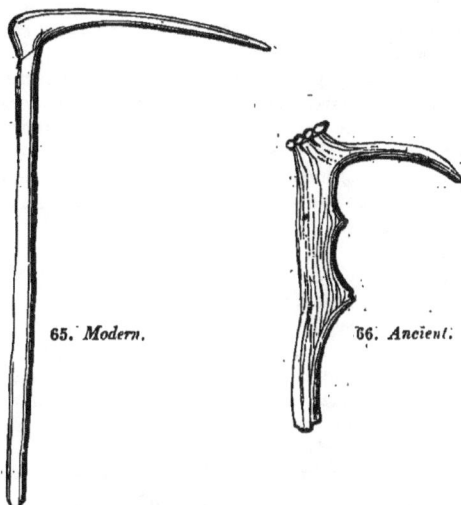

65. *Modern.*   66. *Ancient.*

Figs. 65 and 66.—*Flint Digger's Picks.*

scale. The Grime's Graves pick is drawn from a specimen in
which the tine was broken in the middle; and, as I have not
access to a complete tool, the tip is inserted from a broken one
in my possession. I may add that the tip in question is more
curved than usual, and I have seen specimens in which the curvature much more closely approximated to that of the modern
implement.

This tool is instructive in several particulars. A one-sided pick is itself remarkable; and, so far as I can make out, is peculiar to this locality. I have inquired of ironmongers, and have searched the illustrated catalogues of Sheffield tool-makers, without finding any trace of the use of such a tool in any branch of trade. If such exist, they must be of as local a nature as the one in question.* A one-sided pick possesses no advantage whatever over a two-sided tool *as a pick*, it is peculiar to this district, so that we may justly infer that if it were of comparatively recent introduction there is no reason why a two-sided tool should not have been adopted. It might be suggested that they are more convenient for use in narrow burrows than the larger implement would be; but, as a matter of fact, this is not the case, for the burrows are sufficiently wide for the wielding of an ordinary pick. When we find this singular implement in use just in that particular industry in which a natural one-sided pick was used in early times, we must conclude that it is indeed a relic of the past. The prong is made longer than before, but it preserves much the same thickness.

The natural deer-horn pick has a curvature in the handle which is not ill-adapted for convenient use. Just in the same manner, the most prized modern flint-picks have a double curvature, though it is very slight.

The natural pick was, however, used also as a hammer, for gently tapping the flint in order to loosen it after it has been cracked. This is very patent upon some of the specimens. The picks, for the most part, were made from shed antlers, and the burr around the crown was very nicely adapted for the purpose of a hammer-head, and this is always very much battered, and frequently entirely worn away. The modern flint-pick is also used as a hammer; and, as shown in the figure, the part corresponding with the burr curves slightly outward, and is thickened to strengthen it.

When, then, we see in one simple implement three such peculiarities as the single tine, the thickened butt, and the curved handle; and when we find these characters common to a deer-antler, and know deer-antlers were used as flint-picks formerly, and that such picks are so excessively local, the conviction I have expressed becomes a certainty, and we may assert as a demonstrated fact, that the Brandon flint-knappers are the direct descendants of the neolithic flint-workers.

Passing now to a consideration of the tools used in this singular industry, we find in one of them another coincidence utterly inexplicable on any other supposition than the one I proffer. This is in the case of a flaking-hammer, and it is highly significant that this is the only tool that requires special features to ensure its adaptability to the work performed with it. The quartering-

---

* In Sheffield nearly all, perhaps all, kinds of tools used in every branch of British industry can be obtained, except those belonging to the flint-trade. These are all made in Brandon.

hammer for breaking the stone may be of any shape so long as it is heavy enough. The knapping-hammer merely requires to be light, hard, and elastic, and to have a cutting edge. But the flaking-hammer is a tool *sui generis*, and must possess features which at once distinguish it from all other hammers whatever.

Flaking, as I have shown, is a most difficult art, and the tool with which it is performed must possess the maximum of strength, combined with the minimum of size and striking-surface. A hammer-stroke, to dislodge a flake, must be of a certain strength, and of a rebounding or elastic nature. Moreover, it must be delivered with unerring precision, and the area of impact must be small. The force of the blow is hardly ever greater than is acquired by the natural fall of the hammer from a height of a few inches.

The first flaking-hammers in neolithic, and presumably in palæolithic times also, were ovoid smooth pebbles of quartz or some such tough material. They were grasped in the hand, and were used *au naturel*. The next step in the development was to slightly notch one, or both, sides of the pebble, that it might be held the easier. Afterwards, the stone was trimmed to a more accurate shape, and the notches were cut from either side more symmetrically. The culmination of the development took place when the hammer was accurately shaped and ground, and the notches drilled (nearly always from each side) right through the tool, and a slight handle inserted. I have specimens of the first two stages, and Dr. J. Evans, F.R.S., illustrates the two last in his work on "Ancient Stone Implements."

Now the character of this socket-hole, or *eye*, at once distinguishes a flaking-hammer from any other. In an ordinary hammer the handle is used as a lever, in order to intensify the blow, and as the object of such a hammer is to deliver as strong a blow as possible, the handle has to be made stout, and the eye large. But the handle of a flaking-hammer serves a very different purpose; it is simply a means of delivering the blow with precision; it is, in fact, a guide-rod. As a heavy blow is never delivered with a flaking-hammer there is no necessity for the eye and handle to be large, and they are, in consequence, made as small as possible, in order to ensure as much weight as possible in the head. A heavy blow struck with a flaking-hammer would break the handle.

I have shown, in the preceding memoir, that many neolithic implements possess this very peculiarity of small eyes; and that this has much puzzled archæologists. We now see that the peculiarity in question belongs to a flaking-hammer, and to no other tool. The neolithic bored flaking-hammers are merely developments of the rounded pebble; and the old English flaking-hammer of the Brandon flint-knappers is identical in shape with the stone tool. It has already been shown that this interesting tool was once exclusively used by the flakers, but has given place to the more modern French hammer, and is now nearly extinct.

Passing now to the final process, the knapping, we again meet with connecting links between neolithic and modern times. I have already expressed my opinion that the manufacture of flint implements at Brandon was kept alive in the interval between the decadence of the use of stone as weapons, and its re-introduction in gun-flints, by the constant and unbroken demand for strike-a-lights. Also, I have pointed out the high probability of the truth of Dr. Evans's sagacious suggestion that many of the so-called *scrapers* are, in fact, strike-a-lights; and in the four following figures I reduce this supposition to a certainty; for we have here

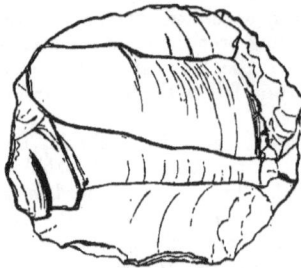

Fig. 67.—*Neolithic Oval Strike-a-Light.*

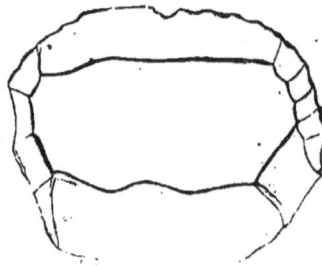

Fig. 68.—*Modern Oval Strike-a-Light.*

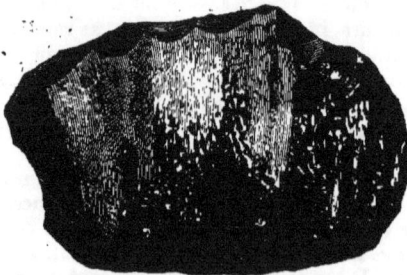

Fig. 69.— *Common English Strike-a-Light.*

the Horse shoe Strike-a-Light and the Oval Strike-a-Light, Fig. 68, two forms of implements, Fig. 67, engraved from a neolithic specimen found by myself, and placed side-by-side with precisely identical specimens which I saw made at Brandon for strike-a-lights, Figs.

Fig. 70.—*Horse-shoe Strike-a-Light.*

Fig. 71.—*Strike-a-Light.*

67 and 70. We may be certain, then, that many so-called *scrapers* are in reality strike-a-lights.

But this remarkable analogy teaches us a more important lesson than the identification of the use of an ancient implement. It shows us that the present Brandon *flint-knappers are making the same kind of implements that the Neoliths made.* This cannot be a mere accidental coincidence ; it must be a relic of the past.

The gun-flint is a development of the strike-a-light, but in rather a peculiar way. It will be noticed that the "ribs" on the back of a strike-a-light run lengthwise down the implement, whereas in the gun-flint they run across. This is a "generic" difference. Now the gun-flints are not modifications of ribbed strike-a-lights at all, but of the ribless kinds made from English flakes (see Fig. 58, Gun-flint Memoir). The earliest gun-flints, therefore, had no ribs. They were rounded at the heel

' F

like strike-a-lights, which they resembled in every respect, being
merely smaller. The French gun-flints still preserve this
character. A perfect gradation can be seen between ribless
strike-a-lights, through ribless gun-flints, and single-backed gun-
flints to the finest modern gun-flints.

We have thus traced a certain community of ideas running
through the neolithic and modern manufactures of flint implements,
which cannot be ascribed to anything but a community of origin.
If the Brandon flint-knappers had re-invented the art, it is ex-
ceedingly unlikely that they would have hit upon all the points
discussed in this paper : the evidence in favour of my supposition
is cumulative, and seems to me irresistible.

As if to. place matters beyond the shadow of doubt, we are
enabled to contrast a really modern flint-art with the Brandon
manufacture. In France the earliest gun-flint factory was started
in 1719. Now we find that the French *cailloteurs* do not work
in the same way as the Neoliths or Brandon men. Their manner
of digging flint, their tools, and their manufactured articles are all
different and attest their modern origin. This is best shown by
grouping the facts in parallel columns as under :—

Neolithic.	Brandon.	French.
1. Worked a number of pits close together.	1. Do the same -	
2. Sank direct to the floor-stone.	2. Do the same -	
3. Drove burrows into the chalk.	3. Do the same	
4. "Drew" the flint in semi-circular spaces.	4. Do the same	Work quite differently.
5. Burrowed about 12 yards -	5. Do the same -	
6. Filled the worked burrows with chalk.	6. Do the same	
7. Used a one-sided pick -	7. Do the same -	
8. Flaked with a round-headed hammer.	8. Did the same -	8. Use a square-headed hammer.
9. Made oval and horse-shoe strike-a-lights.	9. Do the same -	9. Do not make such implements.
10. Under cut the sides of their implements.	10. Do the same -	10. Never do this.

INDEX.

A.

Ancient Flint Pits, 39, ᶜ 9 *et seq.*
Arrow Heads, 40.
Ashley, Mr. H., 11.

B.

Beckmann's "History of Inventions," quoted, 1, 2.
Beer Head, Devon, flint-knapping at, 14.
Best Carbine, 52.
—— Horse Pistol, 56.
—— ——, double-edge, 56.
—— Musket, 48.
—— Upper Crust Flint, 7.
Block, knapping, 19.
Blood Hill, 8.
Boulder Clay, Chalky, age of, 65.
Broomhill, 9; section at, 9; implements at, 10.
Büchse, 2.
Building Flints, 34, 54.
Burrows, flint, 28, 71.

C.

Caillouteurs, 37.
Carbine, flints, 52–55.
Catton, gun-flints at, 10.
Cavenham, 14.
Chalk, descriptions of beds, 7.
——, dip of, 10.
—— heeled Carbine, 55.
—— —— Horse Pistol, 59.
—— —— Musket, 52.
Chalky Boulder Clay, age of, 65.
Cocks, 2.
Collection of Gun-flints described, 45–64.
Common Carbine, 54; double-edge, 54.
—— —— Horse Pistol, 58; double-edge, 58.
—— —— Musket, 50.
Counting Flints, method of, 33.
Cross Piece, 27.

D.

Darbishire, Mr., on Spanish Gun-flints, 15.
Dead Lime, 6.
Development of Fire-arms, 2; of gun-flints, 39, 41.
Digger's, flint, 21.
Dip of Chalk, 10.
Distinctions between natural and artificial features in flint, 43.
Dorling, Mr. S., 30.
Double-edged gun-flints, 46, 49, 53, 54, 56, 57, 58.
Dresden, old gun at, 2.
Drying flint, 27.

E.

Edgways flints, 6.
Elms Plantation, 10.

Elvedon, 13; section at, 13.
—— Lodge, 14.
English Hammer, 18; origin of, 41, 75.
Evans, Dr. J., F.R.S., on Gun-flints, 1; on Palæolithic gravels, 6; on knapping, 32; on strike-a-lights, 39, 43; on surface chipping, 41; on hammer stones, 42, 75.

F.

Fine Double, 62.
—— Pocket-Pistol, 63.
—— Single, 60.
Fire Arms, development of, 2.
Flakes, 30.
Flaking, 28.
—— Candlestick, 21.
—— Hammers, 16; history of, 17, 42, 75.
Fleet Pits, 5.
Flemish wall piece, early, 2.
Flint, at Lingheath, 5; Santon Downham, 8; Broomhill, 9; Shaker's Lodge, 10; Elms Plantation, 10; Norwich, 10; Catton, 10; Icklingham, 11; Elvedon, 13; Elvedon Lodge, 14; Thetford, 14; Cavenham, 14; Tuddenham, 14; King Manor, 14; Grays, 14; Beer Head, 14; Glasgow, 15; Spain, 15; geological position of, 5; descriptions of varieties, 7; from Boulder Clay, 14; digging, 21–25; value of rough, 25; manufacture of, 27–38; origin of gun-flints, 77.
—— digger's Laws, 22.
—— Locks, 3; introduction of, 3.
—— Mines, modern, 21; ancient, 39, 69.
—— Trade, antiquity of, 43, 69.
Floor Stone, 8; method of digging, 23.
Flower, Mr., on palæolithic gravel, 6.
Fractures in flint, artificial and natural, 43.
French gun flints, 37, 68, 78.
Frewer, Mr., 10.

G.

German harquebus, early, 3.
—— Gun flints, 37, 64.
Glacial Beds, 65.
Gravel, palæolithic, 6, 65.
Greenwell, Canon, F.S.A., on Grime's Graves, 69.
Grey Carbine, 55.
—— Horse Pistol, 59.
—— Musket, 51.
Grime's Graves, 39, 40, *et seq* 69.
Gulls, 8.
Gun flints, origin of in France, 3; duration of, 4; decay of trade, 4; statistics of, 4; manufacture of, 27–34; prices of, 34; French, 37; description of collection, 45–64. *See Flint.*

MEMOIRS OF THE GEOLOGICAL SURVEY.

REPORT on CORNWALL, DEVON, and WEST SOMERSET. By Sir H. T. DE LA BECHE, F.R.S., &c. 8vo. 14s.

FIGURES and DESCRIPTIONS of the PALÆOZOIC FOSSILS iu the above Counties. By PROFESSOR PHILLIPS, F.R.S. 8vo. (Out of print.)

THE MEMOIRS of the GEOLOGICAL SURVEY of GREAT BRITAIN, and of the MUSEUM of ECONOMIC GEOLOGY of LONDON. 8vo. Vol. I. 21s.; Vol. II. (in 2 Parts), 42s.

The GEOLOGY of NORTH WALES. By PROFESSOR RAMSAY, LL.D. With an Appendix, by J. W. SALTER, A.L.S. Price 13s. boards. (Vol. III., Memoirs, &c.) (Out of print.)

The GEOLOGY of the LONDON BASIN. Part I. The Chalk and the Eocene Beds of the Southern and Western Tracks. By W. WHITAKER, B.A. (Parts by H. W. BRISTOW, F.R.S., and T. McK. HUGHES, M.A.) Price 13s. boards. Vol. IV.

BRITISH ORGANIC REMAINS. Decades I. to XIII., with 10 Plates each. MONOGRAPH No. 1. On the Genus Pterygotus. By PROFESSOR HUXLEY, F.R.S., and J. W. SALTER, F.G.S. Royal 4to. 4s. 6d.; or royal 8vo. 2s. 6d. each Decade. MONOGRAPH No. 2. On the Structure of Belemnitidæ. By PROFESSOR HUXLEY, LL.D., &c. 2s. 6d.

MONOGRAPH III. On the CROCODILIAN REMAINS found in the ELGIN SANDSTONES, &c. By PROFESSOR HUXLEY, LL.D., F.R.S. Price, with Plates, 14s. 6d.

CATALOGUE of SPECIMENS in the Museum of Practical Geology, illustrative of the Composition and Manufacture of British Pottery and Porcelain. By Sir HENRY DE LA BECHE, and TRENHAM REEKS, Curator. 8vo. 155 Woodcuts. 2nd Edition, by TRENHAM REEKS and F. W. RUDLER. Price 1s. 6d. in wrapper; 2s. in boards.

A DESCRIPTIVE GUIDE to the MUSEUM of PRACTICAL GEOLOGY, with Notices of the Geological Survey of the United Kingdom, the School of Mines, and the Mining Record Office. By ROBERT HUNT, F.R.S., and F. W. RUDLER. Price 6d. (3rd Edition.)

A DESCRIPTIVE CATALOGUE of the ROCK SPECIMENS in the MUSEUM of PRACTICAL GEOLOGY. By A. C. RAMSAY, F.R.S., H. W. BRISTOW, F.R.S., H. BAUERMAN, and A. GEIKIE, F.G.S. Price 1s. (3rd Edit.)

On the TERTIARY FLUVIO-MARINE FORMATION of the ISLE of WIGHT. By EDWARD FORBES, F.R.S. Illustrated with a Map and Plates of Fossils, Sections, &c. Price 5s.

On the GEOLOGY of the COUNTRY around CHELTENHAM. Illustrating Sheet 44. By E. HULL, A.B. Price 2s. 6d.

On the GEOLOGY of PARTS of WILTSHIRE and GLOUCESTERSHIRE (Sheet 34). By A. C. RAMSAY, F.R.S., F.G.S., W. T. AVELINE, F.G.S., and EDWARD HULL, B.A., F.G.S. Price 8d.

On the GEOLOGY of the SOUTH STAFFORDSHIRE COAL-FIELD. By J. B. JUKES, M.A., F.R.S. (3rd Edit.) 3s. 6d.

On the GEOLOGY of the WARWICKSHIRE COAL-FIELD. By H. H. HOWELL, F.G.S. 1s. 6d.

On the GEOLOGY of the COUNTRY around WOODSTOCK. Illustrating Sheet 45 S.W. By E. HULL, A.B. 1s.

On the GEOLOGY of the COUNTRY around PRESCOT, LANCASHIRE. By EDWARD HULL, A.B., F.G.S. (2nd Edition.) Illustrating Quarter Sheet, No. 80 NW. Price 8d.

On the GEOLOGY of PART of LEICESTERSHIRE. By W. TALBOT AVELINE, F.G.S., and H. H. HOWELL, F.G.S. Illustrating Quarter Sheet, No. 63 S.E. Price 8d.

On the GEOLOGY of PART of NORTHAMPTONSHIRE. Illustrating Sheet 53 S.E. By W. T. AVELINE, F.G.S., and RICHARD TRENCH, B.A., F.G.S. Price 8d.

On the GEOLOGY of the ASHBY-DE-LA-ZOUCH COAL-FIELD. By EDWARD HULL, A.B., F.G.S. Illustrating Sheets 63 N.W. and 71 S.W. Price 3s.

On the GEOLOGY of PARTS of OXFORDSHIRE and BERKSHIRE. By E. HULL, A.B., and W. WHITAKER, B.A. Illustrating Sheet 13. Price 3s. (Out of print.)

On the GEOLOGY of PARTS of NORTHAMPTONSHIRE and WARWICKSHIRE. By W. T. AVELINE, F.G.S. Illustrating Quarter Sheet 53 NE. 8d.

On the GEOLOGY of the COUNTRY around WIGAN. By EDWARD HULL, A.B., F.G.S. Illustrating Sheet 89 S.W. on the One-inch Scale, and Sheets 84, 85, 92, 93, 100, 101 on the Six-inch Scale, Lancashire. (2nd Edition.) Price 1s.

On the GEOLOGY of TRINIDAD (West Indian Surveys). By G. P. WALL and J. G. SAWKINS, F.G.S., with Maps and Sections. 12s.

On the GEOLOGY of JAMAICA (West Indian Surveys). By J. G. SAWKINS, &c. With Maps & Sections. 8vo. 1871. Price 9s.

COUNTRY around ALTRINCHAM, CHESHIRE. By E. HULL, B.A. Illustrating 80 NE. Price 8d.

GEOLOGY of PARTS of NOTTINGHAMSHIRE and DERBYSHIRE. By W. T. AVELINE, F.G.S. Illustrating 82 SE. 8d.

COUNTRY around NOTTINGHAM. By W. T. AVELINE, F.G.S. Illustrating 71 NE. Price 8d.

The GEOLOGY of PARTS of NOTTINGHAMSHIRE, YORKSHIRE, and DERBYSHIRE. Illustrating Sheet 82 NE. By W. TALBOT AVELINE, F.G.S. Price 8d.

The GEOLOGY of SOUTH BERKSHIRE and NORTH HAMPSHIRE. Illustrating Sheet 12. By H. W. BRISTOW and W. WHITAKER. Price 3s.

The GEOLOGY of the ISLE OF WIGHT, from the WEALDEN FORMATION to the HEMPSTEAD BEDS inclusive, with Illustrations, and a List of the Fossils. Illustrating Sheet 10. By H. W. BRISTOW, F.R.S. Price 6s.

The GEOLOGY of EDINBURGH. Illustrating Sheet 32 (Scotland). Price 4s. By H. H. HOWELL and A. GEIKIE.

The GEOLOGY of the COUNTRY around BOLTON, LANCASHIRE. By E. HULL, B.A. Illustrating Sheet 89 S.E. Price 2s.

The GEOLOGY of BERWICK. Illustrating Sheet 34 (Scotland). 1 inch. By A. GEIKIE. Price 2s.

The GEOLOGY of the COUNTRY around OLDHAM. By E. HULL, B.A. Illustrating 88 SW. Price 2s.

The GEOLOGY of PARTS of MIDDLESEX, &c. Illustrating Sheet 7. By W. WHITAKER, B.A. Price 2s.

The GEOLOGY of the COUNTRY around BANBURY, WOODSTOCK and BUCKINGHAM. Sheet 45. By A. H. GREEN, M.A. Price 2s.

The GEOLOGY of the COUNTRY between FOLKESTONE and RYE. By J. DREW, F.G.S. (Sheet 4.) Price 1s.

The GEOLOGY of EAST LOTHIAN, &c. (Maps 30, 34, 41, Scot.) By H. H. HOWELL, F.G.S., A. GEIKIE, F.R.S., and J. YOUNG, M.D. With an Appendix on the Fossils by J. W. SALTER, A.L.S.

The GEOLOGY of part of the YORKSHIRE COAL-FIELD (88 S.E.) By A. H. GREEN, M.A., J. R. DAKYNS, M.A., and J. C. WARD, F.G.S. Oct. 1869. 2s.

The GEOLOGY of the COUNTRY between LIVERPOOL and SOUTHPORT (90 SE.) By C. E. DE RANCE, F.G.S. Oct. 1869. 3d.

The GEOLOGY of the COUNTRY around SOUTHPORT, LYTHAM, and SOUTH SHORE. By C. E. DE RANCE, F.G.S.

The GEOLOGY of the CARBONIFEROUS ROCKS NORTH and EAST of LEEDS, and the PERMIAN and TRIASSIC ROCKS about TADCASTER. By W. T. AVELINE, F.G.S., A. H. GREEN, M.A., J. R. DAKYNS, M.A., J. C. WARD, F.G.S., and R. RUSSELL. 8d.

The GEOLOGY of the NEIGHBOURHOOD of KIRKBY LONSDALE and KENDAL. By W. T. AVELINE, F.G.S., T. McK. HUGHES, M.A., F.S.A., and R. H. TIDDEMAN, B.A. Price 2s.

The GEOLOGY of the NEIGHBOURHOOD of KENDAL, WINDERMERE, SEDBERGH, and TEBAY. By W. T. AVELINE, F.G.S., and T. McK. HUGHES, M.A., F.S.A. Price 1s. 6d.

The GEOLOGY of the NEIGHBOURHOOD of LONDON. By W. WHITAKER, B.A. Price 1s.

The GEOLOGY of the EASTERN END of ESSEX (WALTON NAZE and HARWICH). By W. WHITAKER, B.A., F.G.S. Price 9d.

The GEOLOGY of the EAST SOMERSET and BRISTOL COALFIELDS. By H. B. WOODWARD, F.G.S. Price 18s.

The GEOLOGY of the NORTHERN PART of the ENGLISH LAKE DISTRICT (101 SE.) By J C. WARD, F.G.S.

The SUPERFICIAL DEPOSITS of SOUTH-WEST LANCASHIRE. By C. E. DE RANCE, F.G.S. Price 17s.

THE COAL-FIELDS OF THE UNITED KINGDOM ARE ILLUSTRATED BY THE FOLLOWING PUBLISHED MAPS OF THE GEOLOGICAL SURVEY.

COAL-FIELDS OF UNITED KINGDOM.

(Illustrated by the following Maps.)

Anglesey, 78 (SW).
Bristol and Somerset, 19, 35.
Coalbrook Dale, 61 (NE & SE).
Clee Hill, 53 (NE, NW).
Denbighshire, 74 (NE & SE), 79 (SE).
Derby and Yorkshire, 71 (NW, NE, & SE), 82 (NW & SW), 81 (NE), 87 (NE, SE), 88 (SE)
Flintshire, 79 (NE & SW).
Forest of Dean, 43 (SE & SW).
Forest of Wyre, 61 (SE), 55 (NE).
*Lancashire, 80 (NW), 81 (NW), 89 (SE,NE, NW, & SW), 88 (SW). (For corresponding six-inch Maps,see detailed list.)
*Leicestershire, 71 (SW), 63 (NW).
Newcastle, 105 (NE & SE).
*North Staffordshire, 72 (NW), 72 (SW), 73 (NE), 80 (SE), 81 (SW).
*South Staffordshire. 54 (NW). 62 (SW).
Shrewsbury, 60 (NE), 61 (NW & SW).
*South Wales, 36, 37, 38, 40, 41, 42 (SE, SW).
*Warwickshire, 62 (NE & SE), 63 (NW & SW), 54 (NE), 53 (NW).
Yorkshire, 86, 87 (SW), 93 (SW).

SCOTLAND.
*Edinburgh, 32, 33.   *Haddington, 32, 33.
Fife and Kinross, 40, 41.

IRELAND.
*Kanturk, 174, 175.   *Castlecomer, 128, 137,
*Killenaule (Tipperary), 146.
(For Sections Illustrating these Maps, see detailed list.)
  * With descriptive Memoir.

GEOLOGICAL MAPS.

Scale, six inches to a mile.

The Coalfields of Lancashire, Northumberland, Cumberland, Westmorland, Durham, Yorkshire, Edinburghshire, Haddington, Fifeshire, Renfrewshire, Dumbartonshire, Dumfriesshire, Lanarkshire, Stirlingshire, and Ayrshire are surveyed on a scale of six inches to a mile.

Lancashire.

47. Clitheroe.	89. Rochdale, &c.
48. Colne, Twiston Moor.	92. Bickerstaffe, Skelmersdale.
49. Laneshaw Bridge.	
55. Whalley.	93. Wigan, Up Holland, &c.
56. Haggate. 6s.	94. West Houghton, Hindley, Atherton
57. Winewall.	
61. Preston.	95. Radcliffe, Peel Swinton, &c.
62. Balderstone, &c.	
63. Accrington.	96. Middleton, Prestwich, &c.
64. Burnley.	
65. Shperden Moor. 4s.	97. Oldham, &c.
66. Leyland.	100. Knowsley, Rainford, &c.
70. Blackburn, &c.	10L Billinge, Ashton, &c.
71. Haslingden.	102. Leigh, Lowton.
72. Cliviger, Bacup, &c.	103. Ashley, Eccles.
73. Todmorden. 4s.	104. Manchester, Salford, &c.
77. Chorley.	105. Ashton-under-Lyne.
78. Bolton-le-Moors.	106. Liverpool, &c.
79. Entwistle.	107. Prescott, Huyton, &c.
80. Tottington.	108. St. Helen's, Burton Wood.
81. Wardle. 6s.	
84. Ormskirk, St. John's, &c.	109. Warwick, &c. 6s.
85. Standish, &c.	111. Cheadle, part of Stockport, &c.
86. Adlington, Horwick, &c.	
87. Bolton-le-Moors.	112. Stockport, &c. 4s.
88. Bury Heywood.	113. Part of Liverpool, &c. 4s.

Durham.

Scale, six inches to a mile.

Sheet.	Sheet.
1. Ryton. 4s.	8. Sunderland.
2. Gateshead. 4s.	9. ——— 4s.
3. Jarrow. 4s.	10. Edmond Byers. 4s.
4. S. Shields. 4s.	11. Ebchester.
5. Greenside. 4s.	12. Lontoydy.
6. Winlaton.	13. Chester-le-Street. 6s.
7. Washington.	14. Chester-le-Street.

Durham—cont.

Sheet.	Sheet.
16. Hunstanworth.	25. Wolsingham.
17. Waskerley.	26. Brancepeth.
18. Muggleswick.	32. White Kirkley.
19. Lanchester. 6s. Vertical Section, 39.	33. Hamsterley.
	34. Whitworth.
20. Hetton-le-Hole.	41. Cockfield.
24. Stanhope.	42. Bishop Auckland.

Northumberland.

Scale, six inches to a mile.

47. Coquet Island. 4s.	88. Long Benton.
56. Druridge Bay, &c.	89. Tynemouth.
63. Netherwitton.	92. Haltwhistle.
65. Newbiggin. 4s.	95. Corbridge.
68. Bellingham.	96. Horsley. 4s.
69. Redesdale.	97. Newcastle-on-Tyne. 4s.
72. Bedlington.	98. Walker. 4s.
73. Blyth. 4s.	101.
77. Swinburn.	102. Allendale Town.
78. Ingoe. 6s.	105. Newlands.
80. Cramlington.	107. Allendale.
81. Earsdon.	108. Blanchland.
84. Newborough.	109. Shotleyfield.
85. Chollerton.	110. Wellhope.
86. Matfen.	111. Allenheads.
87. Heddon-on-the-Wall.	

Yorkshire.

100. Limley.	274. Barnsley.
184. Kelbrook.	275. Darfield.
201. Bingley.	276. Brodsworth.
204. Aberford.	281. Langsell.
216. Bradford.	282. Wortley.
217. Calverley.	283. Wath upon Dearne.
218. Leeds.	284. Conisborough.
219. Kippax.	287. Low Bradford.
231. Halifax.	288. Ecclesfield.
232. Biratal.	289. Rotherham.
233. East Ardsley.	290. Braithwell.
234. Castleford.	293. Hallam Moors. 4s.
246. Huddersfield.	295. Handsworth.
260. Honley.	296. Laughton-en-le-Morthen.
272. Holmfirth.	299. ———
273. Penistone.	300. Harthill.

SCOTLAND.

Scale, six inches to a mile.

Edinburghshire.

2. Edinburgh, &c.	12. Penicuik, Coalfields of Lasswade, &c.
3. Portobello, Musselburgh, &c.	13. Temple, &c.
6. Gilmerton, Burdie House, &c.	14. Pathead. 4s.
7. Dalkeith, &c.	17. Brunston Colliery, &c.
8. Preston Hall. 4s.	18. Howgate.

Haddingtonshire;

Six inches to a mile.

8. Prestonpans, &c. Price 4s.	
9. Trenent, Gladsmuir, &c. Price 6s.	
13. Elphinstone, &c. Price 4s.	
14. Ormiston, East Salton, &c.	

Fifeshire.

Six inches to a mile.

24. Markinch, &c.	38. Buckhaven.
25. Scoonie, &c.	35. Dunfermline.
30. Beath, &c.	36. Kinghorn.
31. Auchterderran. 4s.	37. Kinghorn. 4s.
32. Dysart, &c.	

Ayrshire.

Six inches to one mile.

19. Newmilns.	36. Grieve Hill.
26. Glenbuck. 4s.	40. Chiltree.
27. Monkton, &c.	41. Dalleagier.
28. Tarbolton, &c.	42. New Cumnock.
30. Aird's Moss.	46. Dalmellington,
31. Muirkirk. 4s.	47. Benbeock.
33. Ayr, &c.	50. Daily.
34. Coylton.	52. Glenmoat.

MINERAL STATISTICS

Embracing the produce of Tin, Copper, Lead, Silver, Zinc, Iron, Coals, and other Minerals. By ROBERT HUNT, F.R.S., Keeper of Mining Records. From 1853 to 1857, inclusive, 1s. 8d. each. 1858, Part I., 1s. 6d.; Part II., 5s. 1859, 1s. 6d.; 1860, 3s. 6d., 1861, 2s. ; and Appendix, 1s. 1862, 2s. 6d. 1863, 2s. 6d. 1864, 2s. 1865, 2s. 8d. 1866 to 1876 2s. each.

THE IRON ORES OF GREAT BRITAIN.

Part I. The IRON ORES of the North and North Midland Counties of England (Out of print). Part II. The IRON ORES of South Staffordshire. Price 1s. Part III. The IRON ORES of South Wales. Price 1s. 3d. Part IV. The IRON ORES of the Shropshire Coal-field and of North Staffordshire. 1s. 3d.